AGRICULTURE ISSUES AND POLICIES

# PLANT GENETIC RESOURCES AND FOOD SECURITY

# AGRICULTURE ISSUES AND POLICIES

Additional books in this series can be found on Nova's website under the Series tab.

Additional E-books in this series can be found on Nova's website under the E-books tab.

AGRICULTURE ISSUES AND POLICIES

# PLANT GENETIC RESOURCES AND FOOD SECURITY

MARLENA A. DIAZ
EDITOR

Nova Science Publishers, Inc.
*New York*

For permission to use material from this book please contact us:
Telephone 631-231-7269; Fax 631-231-8175
Web Site: http://www.novapublishers.com

## NOTICE TO THE READER



Additional color graphics may be available in the e-book version of this book.

**LIBRARY OF CONGRESS CATALOGING-IN-PUBLICATION DATA**
Plant genetic resources and food security / editor: Marlena A. Diaz.
p. cm.
Includes index.
ISBN 978-1-61761-732-4 (softcover)
1. Germplasm resources, Plant. 2. Crops--Germplasm resources. 3. Food crops--Germplasm resources. 4. Plant genetic engineering. 5. Food security. I. Diaz, Marlena A.
SB123.3.P632 2011
333.95'34--dc22
2010033088

*Published by Nova Science Publishers, Inc. † New York*

# CONTENTS

# PREFACE

Plant genetic resources for food and agriculture (PGRFA) serve as the raw material used by plant breeders and farmers to create new crop varieties. As such, they are viewed by many as the foundation for modern agriculture and as essential for achieving global food security. The United Nations Food and Agriculture Organization estimates that more than three-quarters of the increased crop productivity of the past 30 years is the result of plant breeding, and that future global food security depends to a large extent on the continued improvement of food crops. This book explores the International Treaty on Plant Genetic Resources for Food and Agriculture which provides a general framework for conservation and sustainable use of plant genetic resources

Chapter 1- Plant genetic resources for food and agriculture (PGRFA) serve as the raw material used by plant breeders and farmers to create new crop varieties. As such, they are viewed by many as the foundation for modern agriculture and as essential for achieving global food security. The United Nations Food and Agriculture Organization estimates that more than three-quarters of the increased crop productivity of the past 30 years is the result of plant breeding, and that future global food security depends to a large extent on the continued improvement of food crops. This book explores the International Treaty on Plant Genetic Resources for Food and Agriculture which provides a general framework for conservation and sustainable use of plant genetic resources.

Chapter 2- Crop genetic resources are the basis of agricultural production, and significant economic benefits have resulted from their conservation and use. However, crop genetic resources are largely public goods, so private incentives for genetic resource conservation may fall short of achieving public objectives. Within the U.S. germplasm system, certain crop collections lack sufficient diversity to reduce vulnerability to pests and diseases. Many such genetic resources lie outside the United States. This chapter examines the role of genetic resources, genetic diversity, and efforts to value genetic resources. The report also evaluates economic and institutional factors influencing the flow of genetic resources, including international agreements, and their significance for agricultural research and development in the United States.

Chapter 3- Walk into any grocery store and you will see it for yourself: We are producing an unprecedented bounty of food. Having such an abundant food supply begs the question: Why work so hard at improving our crops and livestock when we are already so successful? The answer is simple. We live in a changing world.

The world's population, now at 6.8 billion people, has more than doubled since the 1950s and is expected to reach 9 billion by 2050. The United Nations Food and Agriculture Organization predicts that food production will need to double by 2050 to meet the increased demand. Water supplies will also be a concern as the need to irrigate crops competes with demands from thirsty cities and suburbs in places as diverse as Beijing, New Delhi, and Phoenix.

Climate change is altering landscapes in ways we are only beginning to understand, affecting air temperatures, rainfall patterns, soil dynamics, and the seasonal cycles so vital for a bountiful harvest. Some experts predict that warmer temperatures will reduce yields and cause global food shortages.

New threats from pests and pathogens are also emerging. Ug99, a fungal pathogen, has become an international threat to wheat supplies since its discovery was reported in Uganda a decade ago. Sheath blight, considered the world's worst rice pathogen, has emerged as more of a danger since the 1970s, when scientists developed higher yielding rice varieties.

Chapter 4- Rice *(Oryza sativa)* is the main dietary staple for more than half the world's population. In 2008, worldwide rice consumption exceeded 430 million metric tons. But the world's continued rice supply is jeopardized by a myriad of factors, including diseases and inability to keep up with demand.

Since the Agricultural Research Service is a world leader in rice research, it's no surprise that scientists at the Dale Bumpers National Rice Research Center in Stuttgart, Arkansas, and the Rice Research Unit in Beaumont, Texas, are involved in domestic and international efforts to improve rice varieties worldwide.

Chapter 5- The common bean—which includes pinto, great northern, navy, black, kidney, and snap beans— is considered by many nutritionists to be a nearly perfect food because of its high protein content and low cost. But it is also susceptible to many diseases that reduce seed and pod quality and yields. Agricultural Research Service scientists from labs across the United States are playing major roles in finding solutions to what ails these legumes.

Chapter 6- Corn is essential to the diets of hundreds of millions of people in developing countries, including those in sub-Saharan Africa But millions of those people are at increased risk of health problems because their corn-based diets lack enough vitamin A. Some 40 million children are afflicted with xerophthalmia, an eye disease that can cause blindness, and 250 million people suffer health problems because of a lack of dietary vitamin A. Agricultural Research Service researchers and colleagues at Purdue University and the International Maize and Wheat Improvement Center (CIMMYT) have made some discoveries that could change that.

Corn contains carotenoids, such as beta-carotene, that our bodies convert to vitamin A, but only a very small percentage of corn varieties have naturally high carotenoid levels. Using genetic and statistical tools, the researchers have identified two genes in corn that are linked to higher beta- carotene levels, and they have developed a cheaper and faster way to screen corn plants for more genes that will produce even higher levels of the essential nutrient. The research is expected to at least triple the levels of carotenoids in Africa's corn and could increase levels in some varieties far beyond that, according to Edward Buckler, a geneticist in ARS's Robert W. Holley Center for Agriculture and Health in Ithaca, New York.

Chapter 7- Potatoes are America's number one vegetable crop. Per capita, Americans consume about 130 pounds annually. Worldwide, it's the fourth largest crop after wheat, rice, and corn. But it's a wonder that the potato makes it to the dinner table at all, given the myriad pests and diseases that can take hold well before harvest.

There's the Columbia root-knot nematode, which costs U.S. growers $20 million annually; the potato tuber moth; and late blight, which caused the Irish Potato Famine of 1845 and is still responsible for significant losses and control expenses today. Chemical fumigants and fungicides have long been a staple defense for these pests and pathogens. But the onset of resistance in new pest or pathogen biotypes—coupled with environmental concerns about long-term pesticide use—has prompted the search for sustainable solutions in the form of genetic resistance.

In: Plant Genetic Resources and Food Security ISBN: 978-1-61761-732-4
Editors: Marlena A. Diaz 

*Chapter 1*

# INTERNATIONAL TREATY ON PLANT GENETIC RESOURCES FOR FOOD AND AGRICULTURE

*Melissa D. Ho*

## SUMMARY

Plant genetic resources for food and agriculture (PGRFA) serve as the raw material used by plant breeders and farmers to create new crop varieties. As such, they are viewed by many as the foundation for modern agriculture and as essential for achieving global food security. The United Nations Food and Agriculture Organization estimates that more than three-quarters of the increased crop productivity of the past 30 years is the result of plant breeding, and that future global food security depends to a large extent on the continued improvement of food crops. This book explores the International Treaty on Plant Genetic Resources for Food and Agriculture which provides a general framework for conservation and sustainable use of plant genetic resources. The International Treaty on Plant Genetic Resources for Food and Agriculture (the Treaty on PGRFA) provides a general framework for conservation and sustainable use of plant genetic resources. The treaty sets up a multilateral system of access and benefit sharing, where all members, in exercise of their sovereignty, provide free (or nearly free) access to each other's plant genetic resources for research, breeding, conservation, and training. The multilateral approach allows members access to germplasm to promote food security and improve crop productivity, lowers transaction costs, and redistributes back to the governing body financial benefits derived from the commercial exploitation of the genetic resources.

Currently, 120 countries are parties to the treaty. The United States signed the treaty on November 1, 2002 (Treaty Doc. 110-19), and it was submitted by the Bush Administration to the Senate for advice and ratification on July 7, 2008. The Senate Foreign Relations Committee heard testimony in support of ratification on November 10, 2009, but to date no further action has been taken. Congress could assess several issues related to ratification of the Treaty on PGRFA, including the implications for the United States' position on the Convention for Biological Diversity; the implications for the United States' position on intellectual property rights; the expectations for future financial commitments under the treaty, especially for capacity-building in developing countries; and the potential implications, if any, for congressional proposals related to international agricultural research and development.

## Importance of Plant Genetic Resources for Food and Agriculture (PGRFA): Background and Issues

Humans depend on plant genetic resources for food and agriculture (PGRFA)[1] for many aspects of survival, including food, fuel, and fiber. A study conducted by the United Nations Food and Agriculture Organization (FAO) concluded that plants contribute the vast proportion of the world's food supply, particularly for developing countries in Africa, Asia, and the Pacific.[2] At a global level, the total dietary energy and protein provided by plants is 84% and 63%, respectively, while animal sources contribute 16% and 37%, respectively (Table 1). Plant resources are even more critical in Africa, where about 93% of food energy and 79% of protein are derived from plant sources. The history of the development and use of PGRFA has been characterized by relatively rapid movements of domesticated crops and animals across and among continents, with ultimately a relatively small number of species representing a very high percentage of the daily diets of people around the world.[3] FAO estimates that four crops—rice, wheat, sugar (beet and cane), and corn—account for over 60% of human calorie intake from plants.

## PGRFA and Global Food Security

Many agricultural scientists and development practitioners believe that PGRFA are the foundation for modern agriculture and are essential for achieving food security. They say that much of the increase in food production over the last half-century can be attributed to innovations achieved through plant breeding, drawing on existing genetic resources.

**Table 1. Summary of Sources of Human Energy and Protein (daily average intake of food energy (kcal) and protein (g))**

| | Food Energy | | Protein | |
|---|---|---|---|---|
| | kcal | % | g | % |
| **World** | | | | |
| Plant Sources | 2,388 | 84.0 | 47.3 | 63.4 |
| Animal Sources | 445 | 16.0 | 27.3 | 36.6 |
| **Africa** | | | | |
| Plant Sources | 2,177 | 92.9 | 45.4 | 79.1 |
| Animal Sources | 167 | 7.1 | 12.0 | 20.9 |
| **Asia and the Pacific** | | | | |
| Plant Sources | 2,343 | 87.2 | 49.3 | 71.0 |
| Animal Sources | 343 | 12.8 | 20.1 | 29.0 |
| **Near East** | | | | |
| Plant Sources | 2,441 | 88.1 | 54.7 | 73.1 |
| Animal Sources | 329 | 11.9 | 20.1 | 26.9 |
| **Europe** | | | | |
| Plant Sources | 2,419 | 72.5 | 46.3 | 46.3 |
| Animal Sources | 916 | 27.5 | 53.6 | 53.7 |
| **Latin America and the Caribbean** | | | | |
| Plant Sources | 2,271 | 81.0 | 38.7 | 52.7 |
| Animal Sources | 534 | 19.0 | 34.7 | 47.3 |
| **North America** | | | | |
| Plant Sources | 2,655 | 72.7 | 42.1 | 37.5 |
| Animal Sources | 998 | 27.3 | 70.2 | 62.5 |

Source: Nutrition Division, Food and Agriculture Organization of the United Nations, Background Study Paper No. 11.

Notes: Food energy is the amount of energy in food that is available through digestion and is expressed above in kilocalories (kcal), where protein is expressed in grams (g).

FAO estimates that more than three-quarters of the increased crop productivity of the past 30 years is the result of plant breeding. FAO and other agricultural experts believe that future global food security depends to a large extent on the continued improvement of food crops—for example, developing new varieties that are higher-yielding, resistant to pests and diseases, resistant to extreme weather events such as drought or flood, and/or regionally adapted to different environments and growing conditions. Crop improvement has also resulted in significant gains in the nutritional value of crop plants. Plant genetic resources serve as the raw material used by plant breeders and farmers to create new crop varieties.

Feeding a growing global population will require a significant increase in food production. Despite a 70% growth in world population, agriculture today provides over 15% more calories per capita than it did 30 years ago. By 2050 the world's population is estimated to reach 9.1 billion, 34% higher than today. About 70% of the world's population will be urban (compared to 49% today), and income levels will be higher than they are today. FAO estimates that farmers will need to increase production by at least 70% by 2050 to satisfy the demand for food due to population growth, urbanization, and rising incomes.[4]

## PGRFA and Interdependency

All countries are interdependent with respect to plant genetic resources for food and agriculture— each relies on others for the genetic basis of its major food crops and for food security. Modern crops and forages have a multitude of parent materials, as exemplified by the development of rice varieties grown all over the world (Table 2). The diets of people around the world have evolved and adapted to such an extent that most countries and regions rely heavily on nonindigenous, imported germplasm of staple crops from other parts of the world. For example, corn is one of the world's three most important staple crops, especially for millions of people in sub-Saharan Africa. Corn originated in South America, but the United States is the largest global producer of corn and holds one of the world's largest genebank collections of corn varieties. Cassava, which also originated in South America, is another major food source in Africa today, while African millets and sorghums are major food crops in South Asia and Latin America. Latin America's extensive cattle pastures depend largely on African grasses. Alfalfa from southwestern Asia is now cultivated around the globe. A plate of pasta with red sauce, a dish typical in Italy, relies on crops that originated in South America (tomatoes) and in west and central Asia (wheat). The exchange of plant genetic resources has taken place over centuries, and without it few typical "local" meals would exist. A recent study concluded that for the major food crops, all regions were dependent on PGRFA from other regions to a high degree—over 50% for most regions.[5] Interdependence in central Africa ranges from 67% to 84%, and in south Asia ranges from 85% to 100%. No country in the study was ranked as completely self-sufficient. The high degree of interdependence argues for continued access by countries to a wide range of plant genetic resources in other regions as essential for crop improvement and the development of modern agriculture.

## PLANT GENETIC RESOURCES AND CROP IMPROVEMENT: THE CASE OF WHEAT VS. CORN

How will the world feed itself in the coming years? Many believe that conventional agriculture will continue to play a critical role, with cereal grains being of primary importance. The International Food Policy Research Institute (IFPRI) has predicted that by the year 2020, almost 96% of the world's rice consumption, two-thirds of the world's wheat consumption and almost 60% of the world's corn consumption will be in developing countries. Forecasts call for wheat to surpass rice in its dominant role in feeding the poor of those nations. It will likely become the most important cereal in the world, with corn close behind.

Wheat is one of the few truly global crops, grown all over the world. It belongs to the genus *Triticum*, which originated almost 10,000 years ago in the Fertile Crescent in central and west Asia (the Middle East). Hundreds of thousands of wild species, landraces, and local cultivars within the *Triticum* species constitute the "wheats" of the world. ("Landrace" refers to domesticated animals or plants adapted to the biological and cultural environment in which they originated or are commonly grown. They often develop naturally or from traditional breeding methods, and are thought to have more diverse characteristics than commercial varieties, allowing them to adapt to more variable and local environments.) Thousands of species of *Triticum* have been collected and are currently stored in genetic resources centers around the world. A study conducted by the International Center for Maize and Wheat Improvement (CIMMYT) found that the number of different landraces in pedigrees of modern wheat varieties has steadily increased during the past 30 years and that the geographical origin of the landraces has broadened. Going beyond rather general and poorly defined contributions to modern varieties, several specific genes that have made major impacts on wheats can be directly traced to contributions from genetic resources. One of the best-known examples of the use of plant genetics for crop improvement is the integration of dwarfing genes (genes that reduce plant height and tillering capacity, which ultimately prevent plant lodging) from a Japanese wheat cultivar to wheat varieties in Mexico by Dr. Norman Borlaug. Dr. Borlaug's work launched the so-called "Green Revolution," which led to higher-yielding wheat varieties, increased food security for millions, especially the poor in Latin America and Asia, and Dr. Borlaug's winning the Nobel Peace Prize in 1970. Other examples include the use of a wild relative of wheat from the eastern Mediterranean to obtain a gene that increases the protein content of bread and durum wheat. Breeders have also called on plant genetic resources from all over the world in continuing efforts to develop disease-resistant wheat varieties to a detrimental pathogen called wheat rust, which has caused severe losses to millions of farmers and threatens wheat production globally, including in the United States and Canada.

Unlike wheat, the use of genetic resources in corn improvement is not well documented globally, and is likely not as widespread. Although approximately 50,000 accessions of corn exist in germplasm banks around the world, most of these have never been adequately evaluated for useful traits. It has been estimated that less than 1% of the U.S. germplasm base is exotic. On a global basis, only around 5% of the available corn germplasm is commercially used. The untapped potential of these genetic resources is indicated to some extent by the progress that U.S. breeders have achieved through plant breeding. Through the development of improved varieties, breeders doubled U.S. corn yields between 1930 and 1966, and tripled 1930 yields by 1995.

At the same time, some in the agricultural community are concerned about the lack of genetic diversity in corn used for crop production. The widespread deployment of genetically uniform varieties increases susceptibility to diseases and pests and does not allow for stable yields in variable environmental conditions. Increases of 1.5%-2.0% per year of genetic gain for yield are still being achieved, but some question whether they can be sustained. The incorporation of exotic germplasm into adapted lines may give rise to additional hybrid vigor and higher yield potential. In addition, several studies have demonstrated that exotic germplasm contains significant variation for many quality traits. Because many of the genetic resources of maize have undergone extensive selection over centuries for indigenous uses such as feed, food, and fodder, a wealth of new qualities and characteristics remains to be discovered. In addition, wild relatives of corn such as teosinte and *Tripsacum* are also viewed as potential sources of novel characteristics.

Sources: David Hoisington, Mireille Khairallah, and Timothy Reeves, et al., "Plant Genetic Resources: What Can They Contribute Toward Increased Crop Productivity?," *Proceedings of the National Academy of Sciences*, vol. 96 (May 1999), pp. 5937-5943. M. Smale, P. Aquino, and J. Crossa, et al., *Understanding Global Trends in the Use of Wheat Diversity and International Flows of Wheat Genetic Resources*, CIMMYT, Economics Working Paper 96-02, Mexico City, 1996.

## U.S. Approach to PGRFA

The U.S. food supply is based on intensive agriculture. Intensive agriculture benefits from genetic uniformity in crops, but it can also increase the potential for crop vulnerability to new pests, diseases, and environmental stresses. An example of that vulnerability occurred in 1970, when a widespread outbreak of a disease called southern corn blight hit from the southeastern United States into the Great Plains. The epidemic cost farmers 15% of the nation's corn crop that year because nearly all the corn planted was genetically susceptible to the fungus that caused the blight. Congress responded to this event by establishing the National Plant Germplasm System (NPGS) within the United States Department of Agriculture's (USDA's) Agricultural Research Service (ARS).[6] The NPGS is a national network of public agencies (federal and state agencies including more than 20 federal gene banks located across the country), private institutions, and individuals. It is the primary entity in the U.S. effort to conserve and use crop germplasm for crop improvement. With a collection that includes about 85 crops, the NPGS collects plant germplasm from all over the world. It is "devoted to the free and unrestricted exchange of germplasm with all nations and permits access to U.S. collections by any person with a valid use,"[7] such as for research or breeding, although medical and other uses are included. Germplasm users in other countries have the same privileges as those in the United States. According to ARS, this policy has "grown out of the belief that germplasm, like the oceans and air, is a world heritage to be freely shared for the benefit of all humanity."[8] Through these efforts, NPGS assists in improving the quality and productivity of crops in the United States and in the world.

**Table 2. Summary of International Flows of Rice Ancestors in Selected Countries**

| Country | Total landrace progenitors in all released varieties | Own landraces | Borrowed landraces |
|---|---|---|---|
| Bangladesh | 233 | 4 | 229 |
| Brazil | 460 | 80 | 380 |
| China | 888 | 157 | 731 |
| India | 3,917 | 1,559 | 2,358 |
| Indonesia | 463 | 43 | 420 |
| Nepal | 142 | 2 | 140 |
| Nigeria | 195 | 15 | 180 |
| Pakistan | 195 | 0 | 195 |
| Philippines | 518 | 34 | 484 |
| Thailand | 154 | 27 | 127 |
| United States | 325 | 219 | 106 |
| Vietnam | 517 | 20 | 497 |

Source: Modified from C. Fowler and T. Hodgkin, "Plant Genetic Resources for Food and Agriculture: Assessing Global Availability," *Annual Review of Environment and Resources*, vol. 29 (November 2004).

Notes: The landrace progenitors listed are for a country's commercially released varieties only; they do not include local landraces grown on a noncommercial basis by farmers.

The Germplasm Resources Information Network (GRIN) provides support to NPGS and gives germplasm users continuous access to databases for the maintenance of passport, characterization, evaluation, inventory, and distribution data important for the effective management and use of national germplasm collections. GRIN is also administered by ARS.[9]

In 1990, Congress authorized the establishment of a National Genetic Resources Program (NGRP). NGRP has the responsibility to acquire, characterize, preserve, document, and distribute to scientists germplasm of all life forms important for food and agricultural production, which, in addition to plants, includes animals, microbes, and invertebrates. The National Genetic Resources Advisory Council (NGRAC) advises and makes recommendations to the Secretary and Director of the NGRP. The NGRAC responds to national issues pertaining to the conservation and utilization of genetic resources for food and agriculture.

## INTERNATIONAL TREATY ON PLANT GENETIC RESOURCES FOR FOOD AND AGRICULTURE

The International Treaty on Plant Genetic Resources for Food and Agriculture (the Treaty on PGRFA) provides a general framework for conservation and sustainable use of plant genetic resources. The treaty's preamble acknowledges that the conservation, exploration, collection, characterization, evaluation, and documentation of PGRFA are essential for sustainable agriculture development and to meet the global goals of ending hunger and poverty, as stated in the Rome Declaration on World Food Security and the World Food Summit Plan of Action.[10] The treaty sets up a multilateral system of access and benefit sharing, where all members, in exercise of their sovereignty, provide free (or nearly free) access to each other's plant genetic resources for research, breeding, conservation, and training. The multilateral approach allows members access to germplasm to promote food security and improve crop productivity, lowers transaction costs, and redistributes back to the governing body financial benefits derived from the commercial exploitation of the genetic resources.[11] The treaty is unlike other international laws governing global genetic resources, such as the Convention on Biological Diversity (CBD; see text box below), which extends private or sovereign control and limitations over genetic resources through bilateral interactions, and which many feel is inappropriate for food and agriculture. By establishing a multilateral approach that provides for a standardized protocol and framework applying to all contracting parties, the treaty deals with access and benefit-sharing of agricultural biodiversity in a different way than they are treated under the CBD.

### History of the Treaty

The treaty originated from and eventually replaced the International Undertaking on Plant Genetic Resources (IU),[12] a voluntary non-legally binding agreement adopted by FAO in 1983.[13] The IU was the first international instrument that sought "to ensure that plant genetic resources of economic and/or social interest, particularly for agriculture, will be explored, preserved, evaluated and made available for plant breeding and scientific purposes."[14] The IU reflected the widely held view of the time that plant genetic resources were a heritage of humanity that should be available to all for research and breeding.

While the IU attracted considerable support,[15] some countries did not find the concept of free availability of genetic resources under the IU compatible with the intellectual property protection afforded by plant breeders' rights.[16] Some tension existed concerning farmers' rights, in that intellectual property regimes that rewarded formal breeders often ignored the contributions of generations of farmers to the development and conservation of the PGRFA that breeders utilize. Many critics were also concerned that any system addressing PGRFA should reflect more fully the sovereign rights that countries have over those resources. These concerns were addressed in a series of agreed interpretations of the IU,[17] adopted in 1989, that sought to balance the rights of breeders and farmers. A further conference resolution in 1991 reiterated the sovereign rights of states over their plant genetic resources.[18]

While negotiations proceeded towards the adoption of the CBD,[19] the parties in an appendix to the Nairobi Final Act of the CBD resolved that there were outstanding issues on the interrelationship between the CBD and the promotion of sustainable agriculture. In 1993, the FAO Conference requested FAO to launch a revision of the IU to take into consideration the outstanding issues of access on mutually agreed terms to PGRFA, including *ex situ* collections[20] and the realization of farmers' rights, in harmony with the CBD, and asked its intergovernmental Commission on Genetic Resources for Food and Agriculture to act as the forum to negotiate between countries.

## Convention on Biological Diversity and the International Treaty on Plant Genetic Resources for Food and Agriculture

The Convention on Biological Diversity (CBD) is a legally binding treaty that was launched at the United Nations Conference on Environment and Development at the Rio "Earth Summit" in 1992. The CBD, which came into full force in December 1993, recognized for the first time in international law that the conservation of biological diversity is "a common concern of humankind" and an integral part of the development process.

The agreement covers all ecosystems, species, and genetic resources (terrestrial and aquatic). It links traditional conservation efforts to the economic goal of using biological resources sustainably. It sets principles for the fair and equitable sharing of benefits arising from the use of genetic resources, notably those destined for commercial use. It also covers the rapidly expanding field of biotechnology through its Cartagena Protocol on Biosafety, addressing technology development and transfer, benefit-sharing, and biosafety issues. The CBD's three primary objectives include (1) conservation of biodiversity; (2) sustainable use of its components; and (3) equitable sharing of the benefits from the utilization of genetic resources. The CBD currently has 191 parties, of which 168 are signatories. President Clinton signed the CBD on behalf of the United States in 1993, but to date it has not received a ratification vote on the Senate floor.

Many believed that the CBD approach, while important for the conservation of biodiversity on earth, was not relevant to plant genetic resources for food and agriculture (PGRFA). The CBD also did not deal with farmers' rights or with pre-existing *ex situ* collections of plant genetic materials stored outside their native habitat (typically through the collection and storage of germplasm in a seedbank or genebank), such as those held by the Consultative Group for International Agriculture Research (CGIAR) Centers and other international organizations. The special nature of PGRFA and the need to seek a special solution for these resources, separate from other genetic resources, was recognized by Resolution 3 of the Nairobi Conference that adopted the CBD in May 1992, by the Conference of Parties to the CBD itself, and in the preamble of the plant treaty. The CBD creates a series of specific commitments related to genetic resources, specifically access and benefit-sharing, typically on a bilateral basis, and its objectives are basically environmentally oriented. The Treaty on PGRFA, by contrast, deals specifically with the conservation and sustainable use of plant genetic resources for food and agriculture on a multilateral basis, and its objectives are more related to food security and agricultural productivity. Under the treaty, PGRFA are exchanged through a standard materials transfer agreement (SMTA) and shared freely for research, breeding, conservation, and training purposes. The treaty essentially carves out a special case within the overall CBD framework and provides for a multilateral approach to PGRFA.

After seven years of complex and difficult negotiations, FAO members concluded the International Treaty on Plant Genetic Resources for Food and Agriculture. The treaty established the legal basis for the exchange of PGRFA, at least for those covered in Annex 1 by the multilateral system of access and benefits. The treaty was adopted by consensus by the 31st session of the FAO Conference on November 3, 2001,[21] and would enter into force 90 days after the ratification, acceptance, approval, or accession of the 40th country, which occurred on June 29, 2004. The treaty currently has 120 contracting parties (see Appendix for list).

The United States signed the treaty on November 1, 2002,[22] and it was submitted by the Bush Administration to the Senate for advice and ratification on July 7, 2008. In her letter of submittal to President Bush, Secretary Rice stated that "[a]ll interested agencies in the Executive Branch favor ratification of the Treaty, which can be implemented under existing authorities." The Senate Foreign Relations Committee heard testimony in support of ratification of the treaty on November 10, 2009, but to date no further action has been taken.[23]

## Summary of the Main Components of the Treaty

### *Treaty Objectives*

The fundamental purpose of the treaty is to enable individuals and nations around the world to make use of plant genetic resources for food and agriculture in order to ensure global food security. The two primary objectives of the treaty, as stated in Article 1, include:

- conservation and sustainable use of plant genetic resources for food and agriculture; and
- fair and equitable sharing of benefits derived from their use, in harmony with the Convention on Biological Diversity, for sustainable agriculture and food security.

### *Summary of Treaty Provisions*

The main components of the treaty are:

- general provisions relating to the conservation and sustainable use of plant genetic resources for food and agriculture;
- farmers' rights,

- the Multilateral System of Access and Benefit Sharing (MLS);
- supporting components;
- financial provisions; and
- institutional provisions.

**Conservation and Sustainable Use of PGRFA**

The general provisions on the conservation and sustainable use of PGRFA apply to all PGRFA, not just those listed in Annex 1 of the treaty. The general provisions set a modern framework for the conservation and sustainable use of PGRFA drawing upon the Global Plan of Action (GPA) for the Conservation and Sustainable Use of PGRFA.[24] Article 5 sets out the main tasks that contracting parties are to carry out with respect to conservation, evaluation, and documentation of PGRFA. Similar to other CBD provisions, the responsibilities are placed on each contracting party, acting individually or, where appropriate, in cooperation with other contracting parties, and call for the promotion of an integrated approach to the exploration, conservation, and sustainable use of PGRFA. Article 6 requires the contracting parties to develop and maintain appropriate policy and legal measures that promote the sustainable use of PGRFA. Articles 7 and 8 deal with national commitments, international cooperation, and technical assistance.

According to analysis provided by the State Department in treaty transmittal documents, the treaty likely could be implemented in the United States under existing policies, programs, and statutory authorities, primarily those under the jurisdiction of USDA. The State Department analysis suggests that the activities described in Articles 5 and 6 are consistent with current U.S. practice and could be implemented using existing USDA authorities to operate the National Plant Germplasm System (NGPS) and for ARS 's research activities derived from 7 U.S.C. § § 1621-27, 2201, 2204, 3291, and 5841. Activities described in Articles 7 and 8 are also consistent with U.S. practice. The U.S. currently participates in the FAO; USDA provision of technical assistance to further the sustainability of global agriculture is currently provided pursuant to 7 U.S.C. § 3291; and USAID has provided program support for International Agricultural Research Centers and international organizations such as FAO to strengthen national agricultural research systems in developing countries pursuant to authority derived from the Foreign Assistance Act of 1961, as amended, 22 U.S.C. § 2220b. Further, the U.S. Patent and Trademark Office sponsors the Global Intellectual Property Academy,[25] which holds seminars for sponsored participants from developing countries and includes conservation and sustainable use of genetic resources.

## Farmers' Rights

Article 9 of the treaty deals with farmers' rights and recognizes the contributions of local and indigenous communities and farmers to the conservation and development of plant genetic resources as a basis for food and agriculture production. Article 9 places the responsibility for realizing the rights of farmers on national governments. The provisions of Article 9 are neutral with respect to the issue of the right of farmers to save, use, exchange, and sell farm-saved seed, an issue that was hotly contested during the negotiations. The wording in the treaty recognizes implicitly that farmers may have rights under national law and that these should in no way be limited by the provisions in Article 9. The measures that contracting parties should take under Article 9 include the protection and promotion of:

- traditional knowledge relevant to PGRFA;
- rights of farmers to participate equitably in the sharing of benefits arising from the utilization of PGRFA; and
- the right to participate in making decisions at the national level with respect to the conservation and sustainable use of PGRFA.

The United States acknowledges the importance of such recognition and consultation pursuant to various national and state laws, regulations, and orders. USDA has long conveyed extensive nonmonetary benefits to farmers through land-grant universities and extension services authorized under 7 U.S.C. §§ 301 et seq., 322 et seq. and 341 et seq. USDA also provided services specifically to indigenous communities through 7 U.S.C. § 3241 and 20 U.S.C. § 1059d.

## Multilateral System of Access and Benefit Sharing

A key focus of the treaty is the Multilateral System of Access and Benefit Sharing (MLS), which was established both to facilitate access to genetic resources of major food crops and forage species and to share, in a fair and equitable way, the benefits arising from the utilization of these resources, in accordance with multilaterally agreed terms and conditions.

Article 11 specifies the PGRFA covered by the MLS as those that are listed on Annex I, are under the management and control of the parties, and are in the public domain. The list in Annex I covers 35 crops and 29 forages, including many major crops important to the United States for either domestic use or export. Many countries wanted a broad and comprehensive list of crops to be included in the MLS. Others wanted the MLS to start off with a more limited list of the most important crops. In theory, the negotiators agreed on a list of crops chosen according to their importance for food security and their interdependence. In practice, the list set out in Annex I was negotiated in part on the basis of the perceived interests of individual negotiating parties, with some crops important to food security being excluded, such as soybeans and groundnuts (peanuts).[26] Nevertheless, the list does include most of the major food crops, including cereals such as rice, wheat, maize, sorghum, and millets; grain legumes such as beans, peas, lentils, chickpeas, and cowpeas; roots and tubers such as potatoes, sweet potatoes, cassava, and yams; oil crops such as coconut, sunflower, and plants in the mustard family such as cabbage and broccoli; and fruits such as citrus, apple, and banana/plantain. Noticeable absences that would appear to fit the food security and interdependence criteria include soybeans, groundnuts (peanuts), sugar cane, wild relatives of cassava including the genus *Manihot*, several fruits, and tomato.[27] The MLS also includes PGRFA held in the *ex situ* collections of the Consultative Group on International Research (CGIAR) Centers as well as those held in other international institutions, by agreement with the governing body as referenced in Article 15.

The contracting parties are required to take appropriate measures to encourage natural and legal persons in their jurisdictions to include their holdings of Annex I PGRFA in the MLS. The United States currently encourages private entities to deposit germplasm in the National Plant Germplasm System pursuant to authority derived from 7 U.S.C § 5841.

Article 12 creates the core obligation of the treaty, where parties are required to facilitate access to covered PGRFA. Parties are only obliged to provide access to PGRFA under the MLS when the PGRFA will be used solely for the purpose of research, breeding, and training for food and agriculture (not chemical, pharmaceutical, or other non-food or -feed industrial uses). Parties are to provide PGRFA expeditiously and for free or at a minimal charge, and also are to include available passport data for the PGRFA. Article 12 also notes that recipients shall not claim any intellectual property or other rights that limit access to PGRFA or their genetic parts or components, in the form received from the MLS. Recipients are required to continue to make accessed PGRFA available to the MLS under the terms of the treaty. This article also provides for a standard material transfer agreement (SMTA) between germplasm donors and recipients, which is to accompany any transfer of PGRFA under the MLS. The governing body adopted the text of the SMTA in June 2006.

The State Department analysis asserts that the obligations in Article 12 could be implemented in the United States using existing authorities, particularly through ARS, which maintains the National Plant Germplasm System, a network of more than 20 federal gene banks that operate under authority derived from 7 U.S.C. §§ 2201, 2204, 3125a, 3291, 5841, and 5924. Under these authorities, the USDA Secretary is authorized to provide, free of charge, samples of germplasm from the federal genebanks to any requestor, so long as such provision is not inconsistent with other laws or regulations. Also, the State Department analysis suggests that in the United States, any recourse required from contractual disputes arising from the SMTA would be available via existing authorities that allow for recognition and enforcement of arbitral judgments in the Federal Arbitration Act, 9 U.S.C. § 201 et seq.

Article 13 describes the types of benefit-sharing that may result from the provision of access to PGRFA. It recognizes that the provision of PGRFA itself is a major benefit to the world community. Other benefit-sharing takes the form of exchange of information, access to and transfer of technology, capacity-building, and financial benefit-sharing arising from the commercialization of PGRFA. Under the monetary benefit-sharing provision found in Article 13 .2d(ii) and the SMTA adopted in June 2006, a recipient of PGRFA who commercializes a product incorporation material accessed from the MLS is to pay 1.1% of gross sales. Recipients who make such a product available without restriction to others for further research and breeding are encouraged but not required to make such a payment. The parties agree in this article that the benefits go back to the governing body and not to any individual country or entity, and that benefits should flow primarily to farmers in all countries who conserve and sustainably use PGRFA.

Again, the State Department suggests that Article 13 could be implemented using existing USDA authorities derived from 7 U.S.C. § 5841 to operate the National Plant Germplasm System. USDA currently provides technical assistance to further the sustainability of global agriculture pursuant to 7 U.S.C. §3291. USAID provides technical assistance for agriculture development in rural areas pursuant to the Foreign Assistance Act of 1961, as amended via 22 U.S.C. § 2151a.

## Supporting Components

Part V of the treaty deals with supporting components, which are activities that lie outside the institutional structure of the treaty itself but provide essential support for proper implementation of the treaty and its objectives. These include promoting the effective implementation of the Global Plan of

Action, encouragement of international plant genetic resources networks, and development and strengthening of a global information system on PGRFA, including a periodic assessment of the state of the world's PGRFA.

**Financial Provisions**

Part VI of the treaty addresses financial resources. Article 18 states that parties are to implement a funding strategy that will assist in the implementation of the treaty's activities. The objectives of the strategy are to enhance the availability, transparency, efficiency, and effectiveness of the provision of financial resources for the treaty. Financial benefits from the commercialization of PGRFA under the MLS are included in the strategy, as well as finances made available through other mechanisms, funds, and bodies. These provisions state that the governing body may establish targets for funding and that the primary use of the resources are for the implementation of plans and programs under the treaty (e.g., providing resources to strengthen technical capacity and infrastructure to assist developing countries in treaty implementation). Voluntary contributions may be provided by parties and other sources, but the treaty does envisage mandatory payments over time by contracting parties.

**Institutional Provisions**

The treaty establishes a governing body composed of representatives from all contracting parties. The governing body acts as the supreme body for the treaty and provides policy direction and guidance for the implementation of the treaty and the MLS. All decisions of the governing body are taken by consensus, or, if it agrees to do so by consensus, the governing body can use another method of decision making for all matters other than amendments to the treaty and its annexes. The treaty also provides for the appointment of a Secretary of the Governing Body, who is appointed by the Director General of the FAO and is required to have the approval of the governing body.

Amendments to the treaty may be proposed by any contracting party and must be adopted by consensus of the parties present at the session of the governing body. Amendments come into force 90 days after two-thirds of the contracting parties ratify, accept, or approve them and apply only to those parties that have ratified, accepted, or approved them. The treaty provides for a dispute settlement mechanism and contains provisions for third-party mediation when negotiations fail. No reservations may be made to the treaty.

**Table 3. Summary of the Main Components of the International Treaty on Plant Genetic Resources for Food and Agriculture**

| Part | Article | Main provisions |
|---|---|---|
| **I Introduction** | **1** | **Objectives:** Establishes that the objectives are the conservation and sustainable use of PGRFA and fair and equitable sharing of benefits arising from their use, in harmony with the CBD, for sustainable agriculture and food security. |
| | **2** | **Use of Terms:** Defines some key terms including "plant genetic resources for food and agriculture" (PGRFA) and "genetic material." PGRFA means any genetic material of plant origin of actual or potential value for food and agriculture; genetic material means any material of plant origin, including reproductive and vegetative propagating material, containing functional heredity. |
| | **3** | **Scope:** Establishes the scope of the treaty to apply to all PGRFA, not just those listed in Annex I to the treaty. |
| **II General Provisions on Conser-vation and Sustainable Utilization of PGRFA** | **4** | **General Obligations:** Requires parties to make sure their laws conform to their treaty obligations. |
| | **5** | **Conservation, Exploration, Collection, Charact-erization, Evaluation, and Documentation of PGRFA:** Lists the main tasks for contracting parties regarding PGRFA and calls for the promotion of an integrated approach to the exploration, conservation, and sustainable use of PGRFA. |
| | **6** | **Sustainable Use of Plant Genetic Resources:** Req-uires contracting parties to develop and maintain appropriate policy and legal measures that promote the sustainable use of PGRFA and gives a non-exhaustive list of the types of measures that may be included. |
| | **7** | **National Commitments and International Cooperation:** Requires contracting parties, where appropriate, to cooperate with other contracting parties and other relevant international organizations in the conservation and sustainable use of PGRFA. |
| | **8** | **Technical Assistance:** Promotes technical assistance to contracting parties, especially those that are developing countries. |
| **III Farmers' Rights** | **9** | **Farmers' Rights:** Recognizes farmers' rights, and the contribution made by farmers and local and indigenous communities to the conservation and development of plant genetic resources, and places the responsibility for realizing those rights on natio-nal governments. Elements include the protection and promotion of (1) traditional knowledge relevant to PGRFA; (2) rights of |

| | | |
|---|---|---|
| | | farmers to participate equitably in the sharing of benefits arising from the utilization of PGRFA; and (3) the right to participate in making decisions at the national level with respect to the conservation and sustainable use of PGRFA. The provision specifically states that "[n]othing in this Article shall be interpreted to limit any rights that farmers have to save, use, exchange and sell farm-saved seed/propagating material, subject to national law and as appropriate." |
| **IV Multilateral System of Access and Benefit Sharing** | **10** | **Multilateral System of Access and Benefit Sharing (MLS):** Recognizes the sovereign rights of nations over their own PGRFA, including that the authority to determine access to those resources rests with national governments. Further recognizes that the contracting parties agree to establish the MLS to facilitate access to PGRFA and to share, in a fair and equitable way, the benefits arising from the utilization of these resources. |
| | **11** | Coverage of the Multilateral Systems: Deals with the coverage of the MLS, specifying that the MLS covers a list of crops set out in Annex I of the treaty and is based on the criteria of their importance for food security and interdependence. |
| | **12** | **Facilitated Access to PGRFA within the Multilateral System:** Contracting parties agree to take the necessary legal or other appropriate measures to provide |
| | **12** | facilitated access through the MLS to other contracting parties and to legal and natural persons under their jurisdiction. Recipients of material through the MLS must not claim intellectual property or other rights that limit facilitated access to PGRFA or their genetic components. Facilitated access is to be accorded through the standard material transfer agreement (SMTA) adopted by the governing body of the treaty. |
| | **13** | **Benefit-Sharing in the Multilateral System:** Sets out the agreed terms for benefit-sharing within the MLS, recognizing that facilitated access to PGRFA itself constitutes a major benefit of the MLS. Other mechanisms for benefit-sharing include the exchange of information, access to and transfer of technology, capacity-building, and the sharing of benefits arising from commercialization. |

**Table 3 (Continued)**

| | | |
|---|---|---|
| **V Supporting Components** | **14** | **Global Plan of Action**:[a] Promotes the effective implementation of the Global Plan of Action and includes the encouragement of international plant genetic resources networks, and the development and strengthening of a global information system on PGRFA, including a periodic assessment of the state of the world's PGRFA. |
| | **15** | ***Ex situ* Collections of PGRFA held by CGIAR Centers and others:** Deals with plant germplasm collections held by the CGIAR Centers and other international institutions in genebanks. The treaty calls on the CGIAR Centers to sign agreements with the governing bodies to bring their collections under the treaty, where CGIAR Center PGRFA listed in Annex I would be made available as part of the MLS. Non-Annex I materials would be made available according to a material transfer agreement adopted by the governing body previously. |
| | **16** | **International Plant Genetic Resource Networks:** Deals with cooperation with international plant genetic resource networks. |
| | **17** | **Global Information Systems on PGRFA**: Parties agree to establish a global information system to facilitate exchange of globally harmonized information, which is critical for the operation of the MLS and safeguarding of PGRFA. |
| **VI Financial Provisions** | **18** | **Financial Resources:** Parties agree to implement a funding strategy to assist in the implementation of the treaty's activities. The strategy aims to enhance the availability, transparency, efficiency, and effectiveness of the provision of financial resources for the treaty. It will include the financial benefits from the commercialization of plant genetic resources under the MLS, and also funds made available through other international mechanisms. |
| **VII Institutional Provisions** | **19** | **Governing Body:** Establishes a governing body composed of all contracting parties. The governing body is the supreme entity for the treaty and provi-des policy direction and guidance for the impleme-ntation for the treaty, especially the MLS. All deci-sions are taken by consensus, unless, by consensus, another method of decision making is agreed to for all matters other than amendments and annexes. The governing body is expected to maintain regular com-munication with other international organizations, especially the CBD, to reinforce institutional cooperation over genetic resources issues. |

| | | |
|---|---|---|
| | **20** | **Secretary:** Provides for a Secretary of the Gover-ning Body that shall be appointed by the Director-General of FAO with the approval of the governing body. |
| | **21** | **Compliance:** Deals with requiring at its first meeting consideration of cooperative and effective procedures and operational mechanisms to promote compliance with the provisions of the treaty and to address the issues of non-compliance. |
| | **22** | **Settlement of Disputes:** Provides a mechanism for dispute settlement and contains provisions for third-party mediation when negotiations fail. |
| | **23** | **Amendments to the Treaty:** May be proposed by any contracting party, shall be adopted by consensus of the parties present at the session of the governing body, and come into force 90 days after two-thirds of the contracting parties ratify, accept, or approve. |
| | **24** | **Annexes:** Includes Annex I, which lists the crops covered under the MLS; and Annex II, which deals with arbitration and conciliation. |
| | **25-35** | **Final Clauses:** Standard final clauses regarding signature, ratification, accession, entry into force (40 parties required), participation of member organi-zations of FAO (such as the European Community), withdrawal (with written notice, withdrawal shall take effect one year from the date of receipt of notification), termination, depository, and authentic texts. No reservations may be made to this treaty. |

Source: CRS analysis, modified from Michael Halewood and Kent Nnadozie, "Giving Priority to the Commons: The International Treaty on Plant Genetic Resources for Food and Agriculture,“ in *The Future Control of Food*, ed. Geoff Tansey and Tasmin Rajotte (Earthscan, 2008).

*A. The Global Plan of Action (GPA) for the Conservation and Sustainable Utilization of Plant Genetic Resources for Food and Agriculture was adopted by the International Technical Conference on Plant Genetic Resources in June 1996. The GPA is an important element for the Intergovernmental Commission on Genetic Resources for Food and Agriculture, which was established by FAO in 1983, to carry out its mandate. The plan is periodically updated in order to allow for the commission to recommend new priorities and to promote the rationalization and coordination of efforts. The GPA can be found at* http://www.fao.org/ag/AGP/AGPS/Pgrfa/Pdf/GPAENG.PDF.

# ISSUES FOR CONGRESS

## Status of Treaty Implementation

Currently 120 countries are parties to the treaty (see Appendix). The United States signed the treaty on November 1, 2002 (Treaty Doc, 110-19), but no further action was taken on it until it was submitted to the Senate for advice and ratification by the Bush Administration on July 7, 2008. The Senate Foreign Relations Committee held a treaty hearing on November 10, 2009, which included testimony in support of ratification of the plant treaty, but no further action has been taken. [28]

Senior advisors from USDA have participated in the negotiations and subsequently in the governing body sessions, but they have served in observer roles only. Many experts in the field expect that the United States will become a party to the treaty and that the Obama Administration will take a fresh look at it. Several proponents of the treaty assert that the United States might provide needed leadership and resources in the areas of agricultural biodiversity conservation and use, international agricultural research and development, and global food security. Some believe that the United States' presence could help to foster more trust and goodwill between contracting parties, which has taken many years to develop, particularly among developing countries.

Others have suggested that some countries (such as Japan and China) may be more inclined to sign on to the treaty if the United States officially ratifies. In addition, the United States might also be able to provide leadership to resolve some outstanding tensions regarding more comprehensive inclusion of Annex 1 crops covered by the MLS.[29]

Critics claim that there is also a need for more resources and capacity strengthening to assist with treaty implementation and the realization of benefits, which have experienced slow progress since the treaty entered into force, especially for developing countries. Despite having signed on, many developing countries often lack the technical expertise, necessary infrastructure, or required resources to carry out effective implementation of the treaty.

One of the most widely cited accomplishments of the treaty to date is the inclusion of the CGIAR *ex situ* collection of agricultural biodiversity under the multilateral system (see text box below).

The crops listed in Annex 1 to the treaty that are covered by the MLS together contribute to about 80% of the world's total energy food supply. Collectively the CGIAR Centers hold about 600,000 accessions,[30] which account for an estimated 30%-60% of the world's crop diversity.[31]

## THE CGIAR CENTERS UNDER THE TREATY

The Consultative Group on International Agricultural Research (CGIAR) is a strategic alliance of country members, international and regional organizations, and 15 international agricultural research centers that mobilizes science to benefit the poor. The CGIAR was established in 1971 with support from the Ford and Rockefeller Foundations, the World Bank, FAQ, and UNDP in response to the threat of widespread global famine. The CGIAR produces new crop varieties, knowledge, and other products that are made widely available to individuals and organizations working for sustainable agricultural development and global food security and nutrition throughout the world. For more information see http://www.cgiar.org.

Collectively the CGIAR Centers represent the largest concerted effort toward collecting, conserving, and utilizing global agricultural resources to promote global crop improvement and food security. Between them, the CGIAR Centers hold about 600,000 accessions, which account for an estimated 30%-60% of the world's crop diversity. The remaining germplasm are stored in other international, regional, and national gene banks, many of which collaborate closely with the CGIAR Centers. The materials in the CGIAR gene banks include traditional varieties and landraces, non-domesticated species, advanced cultivars, breeding lines, and genetic stocks. These collections are considered valuable to the global community for two main reasons. First, unlike most national and private collections, they are made up largely of farmers' landraces and local varieties, material that is particularly rich in diversity. Second, they are held in trust for the international community. Materials and information about them are available, under specific terms, to anyone who inquires. The CGIAR Centers have agreed not to claim legal ownership or to seek intellectual property rights over the material in their collections. They also agreed to maintain the collections to international standards and to provide samples of in-trust materials and information about the material. The material transfer agreement that accompanies each request for samples binds the recipient to the same terms. From 1980 to 2004, the centers distributed approximately 2.2 million samples and acquired approximately 370,000 accessions.

Article 15 of the treaty called on the centers to bring their collections under the purview of the treaty. Material held by the centers of crops included in Annex 1 of the treaty will be made available in accordance with the MLS. Material collected before June 24, 2004 (the date the treaty came into force), that is not listed in Annex 1 will be made available under the MTA currently used by the centers under the in-trust agreements with FAQ. Material not included in Annex 1 received by the centers after June 29, 2004, will be made available on terms agreed between the center and the country where the material originated. The treaty also provides for contracting parties to give facilitated access to PGRFA of the crops in Annex I of the treaty to the CGIAR Centers that have signed the agreements with the governing body.

The CGIAR Centers have helped to rationalize the *ex situ* conservation of crop diversity around the world. The crop diversity collections managed and studied by the centers are considered by many to be the most important and best documented in the world. Throughout the seven years of treaty negotiation, the centers worked to ensure that the collections they hold in trust would be available to all users for research, breeding, and educational purposes. Both practically and legally, these now form the centerpiece of the multilateral system established by the treaty.

If the United States becomes a party to the treaty, some believe that other countries will expect the United States to contribute greater capacity-building resources for the conservation and use of agricultural biodiversity globally, and for the implementation of the MLS provisions by developing countries. Even though contributions are technically made on a "voluntary" basis, FAO does have an Indicative Scale of Contributions that provides a recommendation for how much countries should contribute. There are a number of expectations and potential global commitments for the United States, but Congress may opt to consider the importance and implications of this treaty relative to other pending international issues and agreements—for example, the Convention on Biological Diversity.

## The Convention on Biological Diversity

Article 1 of the treaty states explicitly that the objectives should be carried out "in harmony with the Convention on Biological Diversity." As discussed in the "History of the Treaty" section, the treaty was adapted from the International Understanding (IU) to meet identified gaps in the CBD process related to agricultural biodiversity. The CBD is the only comprehensive international agreement dedicated to the conservation and sustainable use of biodiversity. Only four nations are not parties to the CBD: Andorra, Iraq, Somalia, and the United States.

After extensive involvement by the United States in the six-year drafting and negotiation phases, President George H. W. Bush declined to sign the treaty when it opened for signatures at the 1992 Rio Earth Summit. In June 1993, President Clinton signed the CBD on behalf of the United States and transmitted the CBD to the Senate for advice and consent along with "seven understandings" to accompany the ratification instrument. He noted that existing federal, state, and local laws and programs were "sufficient to enable any activities necessary to effectively implement our responsibilities under the Convention" and that the "Administration does not intend to disrupt the existing balance of Federal and State authorities through the Convention." The Senate Foreign Relations Committee supported CBD ratification by a 16-3 bipartisan vote, subject to the seven understandings. However, the CBD never received a ratification vote on the Senate floor. The Senate has not revisited CBD ratification for 15 years.

Several environmental groups advocate for the U.S. ratification of the CBD because of their support for biological conservation and protection globally.[32] At the same time, other groups are opposed to ratification of the CBD because of the perceived potential restrictions imposed by the CBD on intellectual property rights and the position of the Cartagena Protocol on Biosafety[33] regarding biotechnology. The Cartagena Protocol, which is an international agreement on biosafety and a supplement to the CBD, claims to protect biological diversity from the potential and perceived risks posed by genetically modified organisms (GMOs). The Cartagena Protocol allows countries to invoke the "precautionary principle"[34] when considering the benefits and risks of new technologies such as biotechnology. For example, it allows countries to ban imports of genetically modified crops if they contend that there is not enough scientific evidence that the product is safe. They can also require exporters to label shipments containing genetically altered commodities such as corn or cotton. The United States has been a strong proponent of the use of biotechnology for agriculture and has argued in international trade venues against the blocking of U.S. commodities by the European Community and others because they contain GMOs.

Congress could assess whether ratification of the Treaty for PGRFA would have any bearing on the United States' position on the CBD. While signing on to one does not require the United States to be a party to the other (e.g., becoming a party to the plant treaty without being a signatory on the CBD does not offer up any known policy contradictions), some maintain that consideration of the context and relationship of both treaties is a prudent approach.

## Intellectual Property Rights

The State Department analysis suggests that the United States provide clarifying language to the governing body regarding some interpretations of the MLS provisions, especially those in Article 12 that describe the terms under which recipients accept the PGRFA. According to Article 12, recipients shall not claim any intellectual property or other rights that limit access to PGRFA or their genetic parts or components, in the form received from the MLS. The State Department analysis suggests notifying the governing body of the following upon deposit of its instrument of ratification: "The United States understands that Article 12.3d shall not be construed in a manner that diminishes the availability or exercise of intellectual property rights under national laws." Commercialization of plant genetic materials is allowable under the treaty, but parties must either provide free access to the material to all contracting parties or return 1.1% of gross sales of the commercialized material back to the governing body's "benefit-sharing" fund.

## Funding Strategy

The treaty requires contracting parties to develop and implement a funding strategy for carrying out the treaty plans, programs, and activities, in particular to assist developing countries in implementing their commitments under the MLS and to build capacity to use and conserve plant genetic resources for food and agriculture. The current goal set by the governing body is to raise $116 million over the next five years. The governing body envisions that funding will come from voluntary contributions by developed country contracting parties, international funds, bodies and organizations such as the Global Crop Diversity Trust (see text box below), multilateral institutions such as the Global Environment Facility and the World Bank, and private organizations. Mandatory and voluntary contributions resulting from the commercialization of crop diversity from the treaty's MLS will also provide funds, for example, 1.1% of gross sales from the commercialized product. In the 2008 farm bill (P.L. 110-246), Congress authorized USAID to contribute $60 million to the Global Crop Diversity Trust from FY2008 to FY2012 to assist in the conservation of genetic diversity in food crops. To date, the United States has contributed $14.5 million to the trust, $2.5 million for operational support and $12.5 million for the trust's endowment. For FY2010, USAID has earmarked $10 million for the trust. A key question that arises is whether there will be increasing pressure for the United States to commit additional funds to the governing body, and if so, how much is appropriate for these causes.

What is less clear is how the funds are to be distributed and used. The benefit-sharing provisions give priority to the sharing of resources and benefits with farmers, especially in developing countries, but the details of how this objective would be implemented is not fully articulated. The governing body established a "Benefit-Sharing Fund," which initially distributed $500,000 to 11 projects in its first biennial cycle (2008-2009). These projects addressed one or more of the following priorities: (1) information exchange, technology transfer and capacity building; (2) managing and conserving plant genetic resources on-farm; and (3) the sustainable use of plant genetic resources. More detail may be sought about the longer term, and about a broader strategy for scaling up the use of treaty funds to support a coordinated, sustainable, and efficient set of programs and activities that promote the conservation and use of PGRFA, especially by farmers in developing countries. More information may also be relevant on how the governing body will (or will not) coordinate with existing international partners, such as the CGIAR Centers and the Global Crop Diversity Trust, in carrying out the benefit-sharing objectives.

## THE GLOBAL CROP DIVERSITY TRUST

The Global Crop Diversity Trust is an independent international organization whose mission is to ensure the conservation and availability of crop diversity for food security worldwide. The trust was established in

2004 through a partnership between the United Nations Food and Agriculture Organization (FAO) and the Consultative Group on International Agricultural Research (CGIAR). For more information, see http://www.croptrust.org.

In 2006, the trust entered into a relationship agreement with the governing body of the Treaty on PGRFA. The agreement recognizes the trust as an "essential element" of the treaty's funding strategy in regard to the *ex situ* conservation and availability of plant genetic resources for food and agriculture. The trust leads an international effort to build a more effective, efficient, and sustainable conservation system for crop diversity by setting regional and crop- specific strategies, identifying key funding priorities, and providing core funding and technical assistance to support the implementation of the treaty, especially by developing countries. The trust has established an endowment, the income from which will be used to support the conservation of distinct and important crop diversity through existing institutions. To date, the trust has secured over $135 million from a wide array of donors, including $14.5 million from the United States, with another $10 million earmarked from USAID for FY2010. The trust's ultimate goal is to raise $260 million. The 2008 farm bill (P.L. 110-246) authorizes USAID to contribute $60 million to the trust's endowment over FY2008-FY2012, subject to appropriations of funds, and provided that the U.S. contribution does not exceed 25 percent of total contributions from all sources. Other major donors include Australia, Canada, Germany, Ireland, Norway, Sweden, Switzerland, United Kingdom, the Grains Research and Development Corporation (Australia), and several private corporations and foundations. A number of developing countries have also provided support, including Ethiopia and India.

The trust is also involved with the government of Norway and the Nordic Gene Bank in the establishment of the Svalbard Global Seed Vault. This facility will provide a safety back-up for existing genebank collections, which are vulnerable to war, civil strife, natural disasters, and even equipment failure and mismanagement. The vault has also been touted as providing a means for restoring agriculture in the event of a global catastrophe of some sort. It is designed to hold 3 million samples of different varieties of agricultural crops (in the form of seed).

## Links to U.S. International Agriculture Research and Development Initiatives

The treaty and its objectives have been promoted extensively by international agricultural researchers and development practitioners as a critical factor for ensuring global food security. Yet several of the global food security initiatives proposed by the Administration[35] and Congress do not directly make the link between agricultural biodiversity conservation and use, and agricultural research, development, and economic growth. What, if any, are the links between these initiatives and the objectives and activities carried out by this treaty? Is there a reason to make the connection between these initiatives?

The Global Food Security Act of 2009 (S. 384 in the Senate and H.R. 3077 in the House) authorizes increased investment in agricultural productivity, infrastructure, science and technology, research, education, and extension for hunger and poverty alleviation. The bill emphasizes the importance of agricultural research in developing countries as the primary means to increasing the productivity of smallholder farmers and seeks to strengthen the use of science and technology for agriculture in countries suffering from chronic food insecurity and poverty. At the same time, no mention is made about how any of these proposed programs might relate to the conservation and use of PGRFA, and the need for technical capacity-building in the development of PGRFA for crop improvement purposes, especially in developing countries.

# Appendix. Parties to the International Treaty on Plant Genetic Resources for Food and Agriculture

## As of February 24, 2010: 120

The following instruments have been deposited on the dates indicated.[36]

| Participant | Signature | Ratification | Acceptance | Approval | Accession |
|---|---|---|---|---|---|
| Afghanistan | | | | | 9/11/2006 |
| Algeria | | | | | 13/12/2002 |
| Angola | 10/10/2002 | 14/3/2006 | | | |

**Appendix (Continued)**

| | | | | | |
|---|---|---|---|---|---|
| Argentina | 10/6/2002 | | | | |
| Armenia | | | | | 20/3/2007 |
| Australia | 10/6/2002 | 12/12/2005 | | | |
| Austria | 6/6/2002 | 4/11/2005 | | | |
| Bangladesh | 17/10/2002 | 14/11/2003 | | | |
| Belgium | 6/6/2002 | 2/10/2007 | | | |
| Benin | | | | | 24/2/2006 |
| Bhutan | 10/6/2002 | 3/9/2003 | | | |
| Brazil | 10/6/2002 | 22/5/2006 | | | |
| Bulgaria | | | | | 29/12/2004 |
| Burkina Faso | 9/11/2001 | 5/12/2006 | | | |
| Burundi | 10/6/2002 | 28/4/2006 | | | |
| Cambodia | 11/6/2002 | | 11/6/2002 | | |
| Cameroon | 3/9/2002 | 19/12/2005 | | | |
| Canada | 10/6/2002 | 10/6/2002 | | | |
| Cape Verde | 16/10/2002 | | | | |
| Central African Republic | 9/11/2001 | 4/8/2003 | | | |
| Chad | 11/6/2002 | | 14/3/2006 | | |
| Chile | 4/11/2002 | | | | |
| Colombia | 30/10/2002 | | | | |
| Congo, Republic of Cook Islands | | | | | 14/9/2004 |
| Costa Rica | 10/6/2002 | 14/11/2006 | | | |
| Côte d'Ivoire | 9/11/2001 | 25/6/2003 | | | |
| Croatia | | | | | 6/8/2009 |
| Cuba | 11/10/2002 | 16/9/2004 | | | |
| Cyprus | 12/6/2002 | 15/9/2003 | | | |
| Czech Republic | | | | | 31/3/2004 |
| Democratic People's Republic of Korea | | | | | 16/07/2003 |
| Democratic Republic of the Congo | | | | | 5/6/2003 |
| Denmark | 6/6/2002 | 31/3/2004 | | | |
| Djibouti | | | | | 8/5/2006 |
| Dominican Republic | 11/6/2002 | | | | |
| Ecuador | | | | | 7/5/2004 |
| Egypt | 29/8/2002 | 31/3/2004 | | | |

| | | | | | |
|---|---|---|---|---|---|
| El Salvador | 10/6/2002 | 9/7/2003 | | | |
| Eritrea | 10/6/2002 | 10/6/2002 | | | |
| Estonia | | | | | 3 1/3/2004 |
| Ethiopia | 12/6/2002 | 18/6/2003 | | | |
| European Community | 6/6/2002 | | | 31/3/2004 | |
| Fiji | | | | | 9/7/2008 |
| Finland | 6/6/2002 | 31/3/2004 | | | |
| France | 6/6/2002 | | | 11/7/2005 | |
| Gabon | 10/6/2002 | 13/11/2006 | | | |
| Ghana | 28/10/2002 | 28/10/2002 | | | |
| Germany | 6/6/2002 | 31/3/2004 | | | |
| Greece | 6/6/2002 | 31/3/2004 | | | |
| Guatemala | 13/6/2002 | 1/2/2006 | | | |
| Guinea | 11/6/2002 | | | 11/6/2002 | |
| Guinea-Bissau | | | | | 1/2/2006 |
| Haiti | 9/11/2001 | | | | |
| Honduras | | | | | 14/1/2004 |
| Hungary | | | | | 4/3/2004 |
| Iceland | | | | | 7/8/2007 |
| India | 10/6/2002 | 10/6/2002 | | | |
| Indonesia | | | | | 10/3/2006 |
| Iran, Islamic Republic of | 4/11/2002 | 28/4/2006 | | | |
| Ireland | 6/6/2002 | 31/3/2004 | | | |
| Italy | 6/6/2002 | 18/5/2004 | | | |
| Jamaica | | | | | 14/3/2006 |
| Jordan | 9/11/2001 | 30/5/2002 | | | |
| Kenya | | | | | 27/5/2003 |
| Kyrgyzstan | | | | | 30/8/2009 |
| Kiribati | | | | | 13/12/2005 |
| Kuwait | | | | | 2/9/2003 |
| Lao People's Democratic Republic | | | | | 14/3/2006 |
| Latvia | | | | | 27/5/2004 |
| Lebanon | 4/11/2002 | 6/5/2004 | | | |
| Lesotho | | | | | 21/11/2005 |
| Liberia | | | | | 25/11/2005 |
| Libyan Arab Jamahiriya | | | | | 12/4/2005 |
| | | | | | |

**Appendix (Continued)**

| | | | | | |
|---|---|---|---|---|---|
| Lithuania | | | | | 21/6/2005 |
| Luxembourg | 6/6/2002 | 31/3/2004 | | | |
| Madagascar | 30/10/2002 | 13/3/2006 | | | |
| Malawi | 10/6/2002 | 4/7/2002 | | | |
| Malaysia | | | | | 5/5/2003 |
| Maldives | | | | | 2/3/2006 |
| Mali | 9/11/2001 | 5/5/2005 | | | |
| Malta | 10/6/2002 | | | | |
| Marshall Islands | 13/6/2002 | | | | |
| Mauritania | | | | | 11/2/2003 |
| Mauritius | | | | | 27/3/2003 |
| Morocco | 27/3/2002 | 14/7/2006 | | | |
| Myanmar | | | | | 4/12/2002 |
| Namibia | 9/11/2001 | 7/10/2004 | | | |
| Netherlands | 6/6/2002 | | 18/11/2005 | | |
| Nicaragua | | | | | 22/1 1/2002 |
| Niger | 11/6/2002 | 27/10/2004 | | | |
| Nigeria | 10/6/2002 | | | | |
| Norway | 12/6/2002 | 3/8/2004 | | | |
| Oman | | | | | 14/7/2004 |
| Pakistan | | | | | 2/9/2003 |
| Palau | | | | | 5/8/2008 |
| Panama | | | | | 13/3/2006 |
| Paraguay | 24/10/2002 | | 3/1/2003 | | |
| Peru | 8/10/2002 | 5/6/2003 | | | |
| Philippines | | | | | 28/9/2006 |
| Poland | | | | | 7/2/2005 |
| Portugal | 6/6/2002 | | | 7/11/2005 | |
| Qatar | | | | | 1/7/2008 |
| Republic of Korea | | | | | 20/1/2009 |
| Republic of Serbia 1 | 1 / 10/2002 | | | | |
| Romania | | | | | 31/5/2005 |
| Saint Lucia | | | | | 16/7/2003 |
| Samoa | | | | | 9/3/2006 |
| Sao Tome and Principe | | | | | 7/4/2006 |
| Saudi Arabia | | | | | 17/10/2005 |
| Senegal | 9/11/2001 | 25/10/2006 | | | |
| Seychelles | | | | | 30/05/2006 |
| Sierra Leone | | | | | 20/11/2002 |
| Slovenia | | | | | 11/1/2006 |

| | | | | | |
|---|---|---|---|---|---|
| Spain | 6/6/2002 | 31/3/2004 | | | |
| Sudan | 10/6/2002 | 10/6/2002 | | | |
| Swaziland | 10/6/2002 | | | | |
| Sweden | 6/6/2002 | 31/3/2004 | | | |
| Switzerland | 28/10/2002 | 22/11/2004 | | | |
| Syrian Arab Republic | 13/6/2002 | 26/8/2003 | | | |
| Thailand | 4/11/2002 | | | | |
| The Former Yugoslav | 10/6/2002 | | | | |
| Republic of Macedonia | | | | | |
| Togo | 4/11/2002 | 23 October 2007 | | | |
| Trinidad and Tobago | | | | | 27/10/2004 |
| Tunisia | 10/6/2002 | 8/6/2004 | | | |
| Turkey | 4/11/2002 | 7/6/2007 | | | |
| Uganda | | | | | 25/3/2003 |
| United Arab Emirates | | | | | 16/2/2004 |
| United Kingdom | 6/6/2002 | 31/3/2004 | | | |
| United Republic of | | | | | 30/4/2004 |
| Tanzania | | | | | |
| United States of America | 1/1 1/2002 | | | | |
| Uruguay | 10/6/2002 | 1/3/2006 | | | |
| Venezuela | 11/2/2002 | 17/5/2005 | | | |
| Yemen | | | | | 1/3/2006 |
| Zambia | 4/11/2002 | 13/3/2006 | | | |
| Zimbabwe | 30/10/2002 | 5/7/2005 | | | |

# End Notes

[1] The International Treaty on Plant Genetic Resources for Food and Agriculture defines PGRFA as "any genetic material of plant origin of actual or potential value for food and agriculture," where "genetic material" is further defined as "any material of plant origin, including reproductive and vegetative propagating material, containing functional heredity."

[2] Nutrition Division (FAO), *Nutritional Value of Some of the Crops Under Discussion in the Development of a Multilateral System*, Food and Agriculture Organization of the United Nations, Background Study Paper No. 11, Rome, Italy, April, 2001, *ftp://ftp.fao.org/docrep/fao/meeting/015/j0748e.pdf*.

[3] Jared Diamond, *Guns, Germs, and Steel: The Fates of Human Societies* (W. W. Norton, 1997).

[4] Food and Agriculture Organization of the United Nations, *How to Feed the World in 2050*, Rome, October 2009, http://www.fao.org/fileadmin/templates/wsfs/docs/expert_paper/How_to_Feed_the_World_in_2050.pdf.

[5] Ximena Flores Palacios, *Contribution to the Estimation of Countries' Interdependence in the Area of Plant Genetic Resources*, Food and Agriculture Organization of the United Nations, Background Study Paper No. 7, Rome, Italy.

[6] For more information, see the National Plant Germplasm System website at *http://www.ars-grin.gov/npgs/.*

[7] Agricultural Research Service, *Seeds for Our Future: The U.S. National Plant Germplasm System*, United States Department of Agriculture, Program Aid 1470, Washington, DC, 1996, *http://sun.ars-grin.gov/npgs/* Seeds_for_Our_Future_Revised_1996.pdf.

[8] Ibid.

[9] For more information, the Germplasm Resources Information Network website is at http://www.ars-grin.gov/.

[10] Announced at the World Food Summit in November 1996, the Rome Declaration resulted in heads of state reaffirming "the right of everyone to have access to safe and nutritious food, consistent with the right to adequate food and the fundamental right of everyone to be free from hunger" and pledging "political will and our common and national commitment to achieving food security for all and to an ongoing effort to eradicate hunger in all countries, with an immediate view to reducing the number of undernourished people to half their present level no later than 2015." For the full declaration text, see http://www.fao.org/docrep/003/w3613e/w3613e00.HTM.

[11] Michael Halewood and Kent Nnadozie, "Giving Priority to the Commons: The International Treaty on Plant Genetic Resources for Food and Agriculture," in *The Future Control of Food*, ed. Geoff Tansey and Tasmin Rajotte (Earthscan, 2008).

[12] See ftp://ftp.fao.org/ag/cgrfa/iu/iutextE.pdf.

[13] Resolution 8/83, ftp://ftp.fao.org/ag/cgrfa/Res/C8-83E.pdf. The IU was overseen by the FAO Commission on Genetic Resources for Food and Agriculture (CGRFA), an intergovernmental body to which 168 countries belong. The CGRFA acted as the interim committee for the treaty, and prepared the first session of the treaty's governing body.

[14] International Undertaking, Article 1.

[15] There were 113 countries that adhered to the International Undertaking.

[16] Plant breeders' rights (PBR), also known as plant variety rights (PVR), are rights granted to breeders of new varieties of plants that give them exclusive control over the propagating material (including seed, cuttings, divisions, tissue culture) and harvested material (cut flowers, fruit, foliage) of a new variety for a number of years. With these rights, the breeder can choose to become the exclusive marketer of the variety, or to license the variety to others. In order to qualify for these exclusive plant breeders' rights, a variety must be new, distinct, uniform, and stable.

[17] FAO Resolutions 4/89 and 5/89.

[18] Conference Resolution 3/91.

[19] The CBD was signed in May 1992 and entered into full force in December 1993.

[20] *Ex situ* collections are plant genetic material that is stored outside of its native habitat, typically through the collection and storage of germplasm in a seedbank or genebank.

[21] The draft treaty was adopted with 116 votes in favor, zero against, and two abstentions by Japan and the United States.

[22] Treaty Document 110-19.

[23] For more information about the treaty ratification process, see CRS Report 98-3 84, *Senate Consideration of Treaties*, by Betsy Palmer.

[24] The Global Plan of Action (GPA) for the Conservation and Sustainable Utilization of Plant Genetic Resources for Food and Agriculture was adopted by the International Technical Conference on Plant Genetic Resources in June 1996. The GPA is an important element for

the intergovernmental Commission on Genetic Resources for Food and Agriculture, which was established by FAO in 1983, to carry out its mandate. The plan is periodically updated in order to allow the commission to recommend new priorities and to promote the rationalization and coordination of efforts. The GPA can be found at http://www.fao.org/ag/AGP/AGPS/Pgrfa/Pdf/GPAENG.PDF.

[25] For more information, see *http://www.uspto.gov/ip/training*

[26] Each country in the negotiations had the opportunity to exclude any crop from the list. In some cases, had countries agreed to include particular crops, this might have sparked reciprocal concessions from other countries on other crops.

[27] Gerald Moore and Witold Tymowski, *Explanatory Guide to the International Treaty on Plant Genetic Resources for Food and Agriculture*, The World Conservation Union (IUCN), IUCN Environmental Policy and Law Paper No. 57, Cambridge, UK, 2005, http://www.nature-worldwide.info/downloads/iucn/guide-treaty-plant-genetic-resources

[28] Transcripts of the hearing testimony are available at the Senate Foreign Relations Committee website at http://foreign.senate.gov/hearings/hearing/20091110/.

[29] Currently soybeans, groundnuts, tomatoes, citrus, Manihot, and some other important food crops are not listed in Annex 1 owing to ongoing disputes between China, countries in Latin America, and others.

[30] Distinct varieties of plants.

[31] Based on the FAO estimate that there are approximately 1 million to 2 million unique accessions globally. David Hoisington, Mireille Khairallah, and Timothy Reeves, et al., "Plant Genetic Resources: What Can They Contribute Toward Increased Crop Productivity?," *Proceedings of the National Academy of Sciences*, vol. 96 (May 1999), pp. 5937-5943.

[32] E.g., Defenders of Wildlife, Center for Biological Diversity, and Society for Conservation Biology; see
http://www.defenders.org/resources
the_u.s._and_the_convention_on_biological_diversity.pdf?ht=.

[33] Full text of the Cartagena Protocol can be found at http://www.cbd.int/doc/legal/cartagena-protocol-en.pdf.

[34] The precautionary principle, as generally defined, states that if an action or policy might cause severe or irreversible harm to the public or to the environment, in the absence of a scientific consensus that harm would not ensue, the burden of proof falls on those who would advocate taking the action. The United States has opposed using it as a binding legal principle.

[35] For more about the Administration's Global Food Security Initiative, see CRS Report R40945, *The U.S. Global Food Security Initiative: Issues for Congress*, by Charles E. Hanrahan and Melissa D. Ho.

[36] Dates are given in the format day/month/year.

In: Plant Genetic Resources and Food Security ISBN: 978-1-61761-732-4
Editors: Marlena A. Diaz 

*Chapter 2*

# CROP GENETIC RESOURCES: AN ECONOMIC APPRAISAL

***Kelly Day Rubenstein, Paul Heisey, Robbin Shoemaker, John Sullivan and George Frisvold***

## ABSTRACT

Crop genetic resources are the basis of agricultural production, and significant economic benefits have resulted from their conservation and use. However, crop genetic resources are largely public goods, so private incentives for genetic resource conservation may fall short of achieving public objectives. Within the U.S. germplasm system, certain crop collections lack sufficient diversity to reduce vulnerability to pests and diseases. Many such genetic resources lie outside the United States. This chapter examines the role of genetic resources, genetic diversity, and efforts to value genetic resources. The report also evaluates economic and institutional factors influencing the flow of genetic resources, including international agreements, and their significance for agricultural research and development in the United States.

## SUMMARY

Half the yield gains in major U.S. cereal crops since the 1930s are attributed to genetic improvements. Demand for crops continues to grow, and environmental conditions change, so continued productivity growth—and the genetic diversity that helps sustain it—remains important. Genetic diversity can be conserved in farmers' fields, in ecosystems that contain wild relatives of cultivated varieties, and in national or international germplasm collections. It is difficult to determine the best mix of conservation strategies. Regardless, the use of genetic resources by one farmer or plant breeder does not preclude their use by another, so private incentives to sustain diverse genetic resources are low. This motivates public measures (and underlying research) to conserve genetic resources.

### What Is the Issue?

Crop genetic resources are the basis from which all crop production stems. But habitat loss, the dominance of scientifically bred over farmer-developed varieties, and genetic uniformity are all threats to continued diversity. Plant breeders need diverse germplasm to sustain productivity growth. The U.S. system for genetic resource conservation may lack sufficient diversity to reduce some crops' vulnerability to pests and diseases. The genetic uniformity of many modern crop varieties has also raised concerns that crop yields and production will become more vulnerable to evolving pests and diseases. At the same time, genetic resource conservation is expensive, and both private incentives and public funding are limited.

Many sources of diverse genetic resources lie outside the United States. To slow or prevent loss of crop genetic diversity worldwide, international agreements have been designed to encourage preservation of genetic diversity and promote the exchange of germplasm. For example, the new International Treaty on Plant Genetic Resources for Food and Agriculture will govern the exchange of germplasm for crops like wheat, maize, and cotton. But implementation has been hampered by a lack of consensus among the treaty's parties on the value of particular genetic resources. Thus, many of the treaty's provisions, such as procedures for transferring germplasm, are still vague. U.S. policymakers and genetic resource managers will face new exchange terms and rules governing the sharing of benefits from commercialized products among the treaty's parties, so the time is right to examine of the costs and benefits of conserving genetic resources.

## What Did the Study Find?

Since crop genetic resources are largely public goods, private returns to the holders of crop genetic resources are lower than their values to the world. Thus, private incentives for conservation are likely not sufficient to achieve a level of crop genetic diversity that is socially optimal. Significant economic benefits derive from conserving and using genetic resources. For example, a one-time, permanent yield increase from genetic improvements for five major U.S. crops has generated an estimated $8.1-billion gain in economic welfare worldwide. The estimated stream of benefits from genetic enhancement activities exceeds the cost of investments in genetic resource preservation and use. Consumers in both the developed and developing world have benefited from higher yields and lower world prices for food. Without continued genetic enhancement using diverse germplasm from both wild and modified sources, the gains in crop yields obtained over the past seven decades are not sustainable, and yields might eventually grow more slowly (or even decline). Agricultural production increasingly relies on "temporal diversity," changing varieties more frequently to maintain resistance to pests and diseases.

Three factors contribute to loss of genetic diversity—habitat loss, conversion from landraces (farmer-developed varieties) to scientifically bred varieties, and genetic uniformity in scientifically bred varieties. The loss of wild relatives occurs mainly through habitat conversion for agricultural use. Habitat loss is particularly problematic in developing countries, which often face greater pressures for wild land conversion than do developed countries. Crop genetic diversity also has diminished as landraces are displaced by scientifically developed varieties. Studies show that far less area is planted to landraces worldwide than a century ago. Finally, crop genetic diversity may decline with reductions in total numbers of varieties, concentration of area planted in a few favored varieties, or reductions in the "genetic distance" between these varieties. Thus far, yields for many major crops have been relatively stable as a result, at least in part, of frequent changes in modern varieties and breeders' continued access to diverse genetic resources.

This economic assessment suggests that crop genetic resources are essential to maintaining and improving agricultural productivity. However, a General Accounting Office (1997) study found that current conservation efforts may fall short of what scientists believe are necessary levels for future crop breeding needs, suggesting a role for public policy. Policy initiatives include broad-based programs of multilateral and bilateral financial assistance, stronger intellectual property rights, and international agreements for germplasm exchange. But institutional constraints may prevent these initiatives from achieving their stated goals.

## How Was the Study Conducted?

This chapter examines the role of genetic resources and genetic diversity in agricultural production, and efforts to value genetic resources. From a review of published literature, the report addresses the value of genetic improvements over time and among regions of the world. Given the role of genetic diversity in minimizing pest and disease epidemics, the report explores how incentives for land conservation, the breeding process, and access to modern varieties can affect diversity in the field.

The report also evaluates economic and institutional factors influencing the flow of genetic resources—including international agreements such as the Convention on Biological Diversity and the International Treaty on Plant Genetic Resources for Food and Agriculture—and their significance for agricultural research and development in the United States. This chapter synthesizes existing literature to review three proposed policy tools to conserve plant genetic resources: (1) public investments in genetic resource preservation in their natural settings (*in situ* conservation) and of genetic resources saved in gene banks (*ex situ* conservation); (2) stronger intellectual property rights over genetic inventions, particularly in developing countries; and (3) agreements for transferring genetic materials among countries.

## INTRODUCTION

Genetic resources provide the fundamental mechanics that enable plants to convert soil, water and sunlight into something of critical value to humans—food. Diverse genetic resources allow humans to select and breed plants and animals with desired characteristics, thus increasing agricultural productivity. U.S. agricultural productivity more than doubled over the last century (Ahearn et al., 1998), and much of this productivity increase came from rapidly rising crop yields. Half the yield gains in major U.S. cereal crops since the 1930s are attributed to genetic improvements (OTA, 1987). But demand for agricultural commodities continues to grow, and environmental conditions change, so continued productivity growth—and the genetic diversity that helps sustain it—remains important.

Genetic diversity can be conserved in the form of diverse cultivated varieties in farmers' fields, ecosystems that contain wild relatives of cultivated varieties, and/or germplasm collections that contain samples of wild and cultivated species. Each method is characterized by different costs and benefits, making it difficult to determine the optimal mix of conservation strategies. But each also shares a common feature. The use of genetic resources by one farmer or plant breeder does not generally preclude their use by another, so private incentives to hold and protect genetic resources are generally lower than their value to users as a group or society as a whole. This means that in the absence of appropriate public measures (and underlying research), private efforts to conserve genetic resources are likely to fall short of the conservation levels that are optimal for society.

Previous researchers have contributed to our knowledge about the use and conservation of genetic resources. The National Research Council published a detailed review of the National Plant Germplasm System that included extensive recommendations to improve the system (NRC, 1991). A second, related book presented a broader look at the management of genetic resources (NRC, 1993) and included chapters on economic value and ownership. However, economic methodology has evolved rapidly since this chapter was released, as have the policy instruments that are used to protect and exchange genetic resources.

The Food and Agriculture Organization of the United Nations developed a report based on studies submitted by member countries. *The State of the World's Plant Genetic Resources for Food and Agriculture* (1996b and 1998) was a useful snapshot of genetic resource conservation and technological methods, but provided minimal economic information such as incentive structures or policy tools. In 1997, the U.S. General Accounting Office presented a systematic analysis of the management of the U.S. national genebank system. Recently, the International Food Policy Research Institute published a set of research briefs focused on gene bank valuation. These last two reports focused only on gene banks, and not on all three genetic conservation options.

All these previous reports have been useful, but recent developments in the international exchange of genetic resources call for a concise and current summary of genetic resource conservation in an economic framework. This chapter focuses on our current understanding of the value of genetic resources, trends in genetic diversity (and the economic incentives that affect them), and recent strategies for protecting genetic resources (including the International Treaty on Plant Genetic Resources for Food and Agriculture, which entered into force in June 2004).

## Origins of Crop Genetic Diversity

Human selection of plant varieties for desired traits (such as taste, pest resistance, or seed size) dates from the very beginnings of agriculture. For thousands of years, farmers have selected, saved, and replanted varieties of the crops that humans consume today. "Centers of diversity" developed where intraspecies diversity of crop varieties was particularly high. Most centers of diversity are found where crops were first domesticated, primarily in today's developing countries.

The pace of genetic improvement accelerated with the development of modern breeding techniques that facilitated selection of specific desirable traits. Breeders have crossed different parental material and selected traits to achieve high yields and improved quality for all types of crops. Breeders have also sought resistance to pests, diseases, drought, and other stress. In fact, resistance has become the primary goal of breeding for many crops.

## Definitions

*Biological diversity* refers to the number, variety, and variability among plant, animal, and microorganism species and the ecological systems in which they live. Biological diversity can be defined at three levels. *Genetic diversity* refers to the different genes and variations generally found within a species. The variation among genes across different wheat varieties is an example. *Species diversity* refers to the variety and abundance of different species in a region. Finally, *ecosystem diversity* is exemplified by the variety of habitats, such as grasslands or wetlands, occurring within a region. The term biological diversity can refer to any or all of the three levels of diversity, but in this chapter we will focus particularly on genetic diversity in agricultural crops.

Crop genetic diversity can be conserved in its natural setting (i.e., *in situ*), or it can be collected and conserved outside its natural environment (i.e., *ex situ*). Within the context of crop genetic diversity, there are five basic kinds of genetic resources:

1. **Wild or weedy relatives** are plants that share a common ancestry with a crop species but that have not been domesticated. These can also be a source of resistance traits, but these traits may be difficult to incorporate in final varieties.
2. **Landraces** are varieties of crops improved by farmers over many generations without the use of modern breeding techniques. These varieties are generally very diverse within species, because each is adapted to a specific environment. Within a modern breeding program, they are sometimes used for resistance traits, and extensive efforts are generally required before their genes are usable in a final variety.
3. **Improved germplasm** is any plant material containing one or more traits of interest that has been incorporated by scientific selection or planned crossing.
4. **Advanced (or elite) germplasm** includes "cultivars," or cultivated varieties, suitable for planting by farmers, and advanced breeding material that breeders combine to produce new cultivars.
5. **Genetic stocks** are mutants or other germplasm with chromosomal abnormalities that may be used by plant breeders, often for sophisticated breeding and basic research.

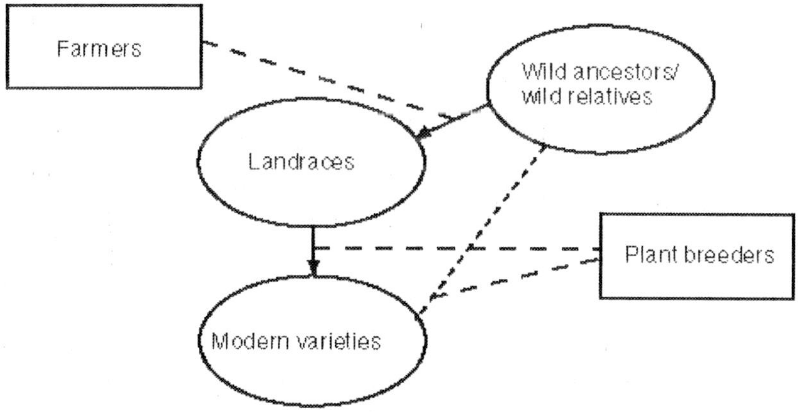


Figure 1. Farmers, plant breeders, and genetic resources

## Current Challenges

Changes in population, income, and other factors (such as urbanization) drive continuing increases in demand for agricultural commodities. Environmental conditions also change and pests and diseases evolve over time, so breeders continually need new and diverse germplasm from outside the utilized breeding stock, sometimes using wild relatives and landraces, to find specific traits to maintain or improve yields (Duvick, 1986). Maintaining resistance is a continual process, because new varieties are resistant to pests and diseases for an average of 5 years, while it generally takes 8 to 11 years to breed new varieties (USDA, 1990).

But private incentives to acquire and preserve genetic resources outside regular breeding stocks are limited, because genetic resources have strong "public goods" characteristics (Brown, 1987; Brown and Swierzbinski, 1985; Frisvold and Condon, 1994; Sedjo, 1992; Simpson and Sedjo, 1992; Reid, 1992; Swanson, 1996). For example, genetic resources are easily transported and replicated, and intellectual property protection has historically been relatively weak for biological innovations, making it difficult for an individual country, firm, or farmer to exclude others from their use. Furthermore, the usefulness of particular genetic resources is highly uncertain, and time horizons for improving genetic resources are long.

Despite the limits on private returns to their conservation and improvement, diverse crop genetic resources remain critical to agricultural production. Therefore, the public sector has played a pivotal role in their conservation. This raises three questions.

First, what are genetic resources worth? Most genetic resources are not market goods; that is, they are not sold as inputs into the breeding process and so lack simple indicators of their value. As such, policymakers find it difficult to compare investment in conservation with other uses for public funds. The international exchange of germplasm is also complicated, as countries may seek to maximize the returns from the set of resources that they hold.

Second, how diverse are genetic resources, not only in gene bank collections but also in the field? Diversity among genetic resources in the field can reduce the prospects for pest and disease epidemics. Farmers generally grow the most productive varieties (in terms of yield or quality), which may or may not be diverse. Society as a whole may prefer a higher level of diversity than farmers do. Incentives for land conservation, the breeding process, and access to modern varieties all can affect diversity in the field. Even the way in which diversity is defined can alter the assessment of benefits associated with different production and conservation decisions.

And finally, what can be done to ensure we have the crop genetic resources that we will need? The reliance of agriculture on these resources suggests the importance of continued preservation efforts. Policy instruments such as funding for *in situ* and *ex situ* conservation, intellectual property rights, and negotiated terms of transfer can be used to promote genetic resource conservation. While these policies can be implemented at the national level, genetic resources are found throughout the world. No nation has all the resources it wants or may need in the future. Thus, international coordination of genetic resource conservation is critical to meeting the longterm requirements of agricultural production.

## ECONOMIC VALUES OF CROP GENETIC RESOURCES

Attaching a value to genetic resources is a complex task. Describing the kinds of benefits associated with these resources is easier. The simplest benefit arises from the direct use of genetic resources: to produce food and fiber or to help create new varieties of crops and livestock. These direct uses are the focus of this chapter, although option value may also be an important motivation for their conservation.[1]

The ultimate direct-use benefits of crop genetic resources are measured in the increased output, higher quality, better resistance to pests, diseases, and other stress, and other characteristics found in improved crop varieties. These benefits derive not only from the genetic resources contained in precursor wild relatives, but also from the efforts of farmers who domesticated the crop and developed landraces through many years of selection; the work of collectors and gene banks that assembled and preserved genetic material in the form of landraces and wild relatives; and the work of plant breeders who have continued to develop and improve crop varieties.

### Estimating the Benefits of Genetic Enhancement

Separating the contributions of breeders from the contributions of the germplasm with which they work is difficult. Thus, many studies have focused on the value of "genetic enhancement," or the value arising from both genetic material and its use by breeders. Most efforts to measure genetic enhancement have focused on specific crop breeding programs, using one of two related methods. The first measures benefits derived from a breeding program directly, and calculates rates of return to plant breeding efforts by comparing breeding program expenditures with their benefits. Many rate-of-return studies depend on the second method, some form of growth accounting. Growth accounting attempts to account for all factors affecting yields and then estimates the portion of the yield increase due to genetic enhancements.[2]

Rate-of-return studies sometimes base their estimates of the benefits from genetic enhancement on experimental estimates of yield gains. Plant breeders and other crop scientists may measure genetic gains in crop yield by conducting experiments that attempt to control for the effects of other inputs.[3] Although these studies focus specifically on genetic gains in yield, they do not always correctly value the economic benefits derived from the use of genetic resources for two reasons. First, yield trials that estimate genetic gains in yield are often conducted with input levels that farmers would not use or under environmental conditions that farmers would not face, in part because such experiments rely on control of other inputs for statistical validity. But plausible farmer responses in the face of changing technologies and market-environmental conditions suggest that yield gains in the field are likely to differ from experimental yield gains (Alston et al., 1995). Second, the resulting supply shifts for individual farmers would need to be aggregated to an industry supply shift in order to analyze economic costs and benefits to all producers and consumers.

Studies valuing the plant breeding component of genetic enhancement (see box, "Economic Studies of the Value of Genetic Enhancement") consistently demonstrate its high utility in creating new varieties with higher yields and better resistance to disease. In most cases, too, the economic benefits of genetic enhancement far surpass the costs. These studies do differ in methodology, so the magnitude of estimated economic benefits is often not consistent across studies. Although Evenson and Gollin (1997) made some efforts to estimate the values of genetic resources directly, for the most part, valuation methodologies have not separated out the contribution made by plant breeding from the contributions of conserving genetic resources in farmers' fields or in gene banks. Nor do most studies provide a detailed welfare analysis of costs and benefits across producers (including non-adopters) and consumers.

## Economic Studies of the Value of Genetic Enhancement

Thirtle (1985) estimated the contributions of biological advances—which include both genetic enhancements and other land-saving technological change—in U.S. crop production using growth accounting (controlling for changes in other inputs such as fertilizers, machinery, and pesticides). Thirtle estimated that biological advances increased corn yields an average of 1.7 percent per year between 1939 and 1978; wheat 1.5 percent; soybeans 1.1 percent; and cotton 0.5 percent. Thirtle further concluded that biological improvements contributed to 50 percent of the yield growth of corn, 85 percent for soybeans, 75 percent for wheat, and 24 percent for cotton. In Thirtle's definition, however, biological improvements included both the use of improved varieties and other land-saving changes in agronomic practices.[1]

Byerlee and Traxler (1995) estimated a rate of return of 52 percent for joint international/national wheat breeding programs in developing countries. Pardey et al. (1996) also used rates of return, focusing on the spillover economic benefits of breeding research—i.e., benefits that accrue in regions or countries other than those originally targeted. They analyzed benefits in the United States (either to U.S. research programs or directly to U.S. farmers) from plant breeding research conducted in 2 of the 15 International Agricultural Research Centers (IARCs) that make up the Consultative Group on International Agricultural Research (CGIAR) system. Pardey et al. estimated returns on U.S. financial support to these two programs and found benefit-cost ratios for the United States of up to 48 to 1 for rice and 190 to 1 for wheat. Brennan et al. (1997) estimated that 64 percent of the genetic improvements to Australian rice came from international germplasm, and that the total Australian benefits of varietal yield improvement from 1962 to 1994 were $848 million (1994).

Evenson and Gollin (1997) estimated that without the International Network for the Genetic Evaluation of Rice, 20 improved varieties of rice would not have been released. The present value of that lost production over a 20-year period (the average length of time a rice variety is economically viable) was estimated to be $1.9 billion. Using a discount rate of 10 percent, the authors estimated that the present value of an added landrace (in a variety introduced by the program) was $50 million.

[1]Technically, Thirtle estimated the rate of land-saving biological-chemical technical change as an exponential time trend within a nested Cobb-Douglas/CES production function. In the same function, a different exponential time trend was used to estimate laborsaving mechanical technological change.

Frisvold et al. (2003) attempted to overcome some of the limitations of earlier studies by adding two features: a global welfare analysis, and a multi-market partial equilibrium model that could calculate the joint effects of genetic improvements in five major crops in the United States between 1975 and 1992.[4] They first estimated the size and distribution of the gross annual benefits of a single-year increase (figure 2, first panel) in the U.S. yields of corn, soybeans, wheat, cotton, and sorghum. About half of the increase in yields can be attributed to improved seed varieties (Fuglie et al, 1996).[5] Accordingly, to simulate the effects of genetic improvement only, the authors increased the supply of crops by half of the average annual yield growth, implicitly assuming no changes in other inputs and no interactions between genetic improvements and other inputs.

Frisvold et al. estimated that the overall economic welfare of U.S. crop producers across the five commodities increased by more than $160 million and that consumer welfare increased by more than $220 million (1989 constant dollars) due to U.S. genetic improvements. Total U.S. economic welfare increased over $350 million. Producers in the rest of the world suffered losses, while consumers in the rest of the world gained from lower world food *prices* global welfare increased by $590 million, with the United States capturing 60 percent of the total gain, other developed countries 25 percent, and developing and transitional economies 16 percent.

In fact, yield increases from genetic improvements are not limited to a single year, so Frisvold et al. also calculated the present value of a permanent increase in yields from genetic improvements (figure 2, second panel).[6] The U.S. benefits of permanent U.S. yield increases range from just under $5 billion (1997 dollars) to over $9 billion. Global benefits range from $8 billion to $15 billion and benefits to developing and transitional economies range from $1 billion to $2.5 billion. (Consumer benefits in developing and transitional economies range from $6 billion to over $11 billion.)

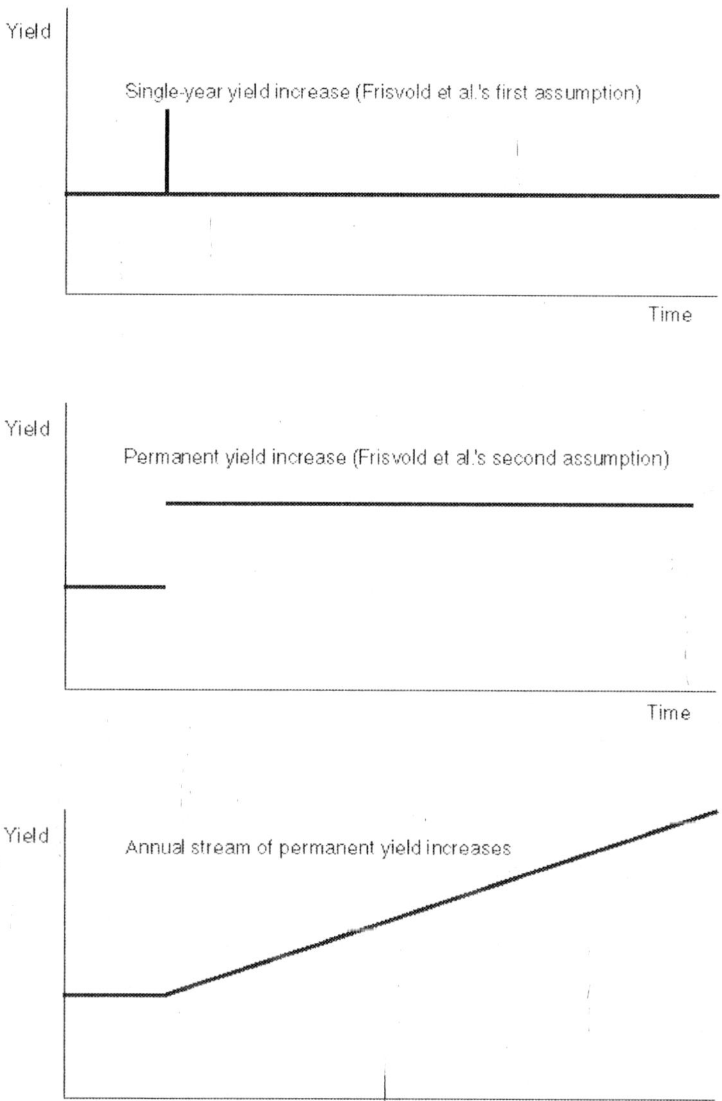


Figure 2. Alternative assumptions about benefits from genetic enhancement

These estimates are conservative for two reasons. First, growth in income and population over time would make the total benefits of yield increases even larger as demand grows. And second, "plant breeding and genetic improvements have not merely generated one-time permanent increases in yields, but rather an annual stream of permanent yield improvements. Every year there is a new incremental permanent increase in yields. The problem is equivalent to receiving a new annuity of varying value every year" (Frisvold et al., 2003) (figure 2, third panel).[7]

These results suggest that investment in genetic enhancement has generated large returns. The United States was the major beneficiary of genetic enhancement in U.S. crops, although the genetic resources used in these improvements might have multiple sources. (Note that Frisvold et al. 's analysis did not include the U.S. research costs necessary to achieve these yield gains.) Nonetheless, developing and transitional economies also benefited from U.S. yield gains, and it is likely that poor consumers in these countries (including many small farmers) are among the major beneficiaries.

## Searching for Valuable Genetic Resources

Genetic enhancement depends on the availability of diverse genetic resources for use by plant breeders. In addition to evaluating genetic enhancement, economists have also attempted to evaluate the search for agricultural genetic resources *in situ* (in their natural habitat), the storage and characterization of these resources *ex situ* (e.g. in germplasm collections), and the search for particular traits within *ex situ* collections. Compared with estimates of returns to genetic enhancement, estimates of search costs and returns often are more complex conceptually and more demanding of scarce data (see box, "Economic Models of Searching for Genetic Resources").

Most models of the economics of searching for genetic resources held *in situ* or *ex situ* have been difficult to apply empirically due to data limitations. Several different types of empirical studies have, however, provided useful information about the economics of conservation. First, Evenson and Gollin (1997) directly estimated likely benefits of additional accessions to the rice collection maintained by the International Network for the Genetic Evaluation of Rice. They estimated that the present value of 1,000 additional accessions (discounted at 10 percent over a 20-year period) was $325 million.

Second, Pardey et al. (2001; 2004) estimated the marginal costs of adding accessions to the *ex situ* gene bank for wheat and maize (corn) at CIMMYT, the International Maize and Wheat Improvement Center, and estimated the cost of holding an additional accession in perpetuity. Though Pardey et al. did not estimate the expected values of benefits for additional accessions (and suggested it might not even be feasible), they argued that the cost of additional wheat accessions was so low that expected benefits would probably always outweigh this cost. They argued that some accessions to the maize gene bank—e.g., landraces and wild relatives—might be more likely to have an expected positive return than others, like recently created breeding lines. This is because useful genetic material contained in breeding lines might well be conserved elsewhere—for example, by maize breeding programs—but useful genetic material in landraces and wild relatives would probably be conserved only in the gene bank.[8]

Third, surveys of plant breeders and other users of gene banks have consistently showed that they find gene bank materials useful. For example, the U.S. National Plant Germplasm System (NPGS) is one of the largest national gene banks in the world; it distributes, for free, more germplasm samples internationally than any other supplier, including the international research centers of the CGIAR. Smale and Day-Rubenstein (2002) found that international users of the NPGS requested materials for a variety of uses, including basic research and breeding, and a majority expected that their use of NPGS materials would stay the same or increase in the future. Of NPGS samples distributed from 1995 through 1999, 11 percent had been used in breeding programs, 18 percent were found useful in other ways, and 43 percent were still being evaluated. Twenty-eight percent of the samples were not considered useful. Rejesus et al. (1996) found that wheat breeders around the world used released cultivars, advanced materials, and germplasm from international nurseries much more frequently than wild relatives and landraces. Wild relatives and landraces were used particularly in search of specific traits, such as disease resistance, drought resistance, and quality.

## Economic Models of Searching for Genetic Resources

Simpson, Sedjo, and Reid (1996) applied a theoretical model, originally used in labor economics, to biodiversity conservation in the context of a search for species of interest to pharmaceutical research. Modifying this model, Simpson and Sedjo (1998) argued that the value to society of biodiversity prospecting (searching for genetic resources currently held *in situ*) for use in crop improvement programs was likely to be low.

Cooper (1998) approached the question as one of investment in "converting" *in situ* genetic resources into *ex situ* resources under (1) uncertainty concerning the measurement and value of *in situ* genetic resources, and (2) irreversibility since *in situ* resources, once lost, cannot be replaced. Cooper's simulations demonstrated that estimates of mean benefits might not be particularly useful, as the range of potential benefits could be quite large.

Evenson and Lemarié (1998) applied a search model to a two-stage process—first, collecting genetic resources *in situ* and placing them *ex situ,* and second, searching the *ex situ* collection for traits of interest. They showed that the optimal size of a collection depends on the number of traits being sought, and on the distribution of genetic resources across geographic regions.

Gollin, Smale, and Skovmand (2000) developed a theoretical model that characterizes the search for resistance to pests and diseases in *ex situ* collections of wheat genetic resources, and then analyzed data on frequency distributions, disease losses, and search costs. They concluded that "the optimal size of search for traits is highly sensitive to the economic magnitude of the problem, the research time lag, and the probability distribution of the trait." Furthermore, even though subcollections of landraces or wild relatives might be used only on rare occasions, high benefits might result on those occasions. The fact that "gene banks and some categories of accessions"—i.e., certain types of genetic materials held by a gene bank—"are infrequently demanded by crop breeders does not in itself imply that marginal accessions have low value."

Drawing on these earlier studies, Rausser and Small (2000) argued that scientific models that "channel research effort towards leads for which the expected productivity of discoveries is highest" significantly reduce search costs from earlier "brute force" models that assume no prior information can be brought to the search. In contrast to the results of Simpson et al., Rausser and Small's simulations suggest that market-based conservation of genetic resources might be possible in some cases because prior information reduces private search costs so they are lower than expected private benefits from searching.

One final consideration refers not to the economics of plant genetic resource conservation *per se,* but to a related scientific development bearing on economic decisionmaking. This is the potential of modern molecular biology, including genomics, to reduce the search costs for useful traits in conserved material. (Genomics refers to investigations into the structure and function of very large numbers of genes undertaken simultaneously.) At this point, however, it is relatively easy to generate mountains of raw genetic sequence data but difficult to transform these data into useful information (Attwood, 2000). Thus, conserved genetic resources may increase in value as genomics and other molecular techniques lower search costs and the costs of capitalizing on search results, but it is difficult to predict the pace at which this will take place.

The literature on searching for valuable genetic resources is less conclusive than the literature on evaluating the benefits of genetic enhancement. The majority of studies agree that economic benefits from searching for genetic resources either *in situ* or *ex situ* are positive compared with costs. This can be true even if successful searches are a small fraction of the total searches conducted. However, studies also conclude that it is quite difficult to value searches for genetic resources, and that the range of potential values may be large. The key variable is information. Application of prior information about the probability distribution of a desired trait or set of traits and where searches are likely to have the highest payoffs can significantly increase the economic value of a search for genetic resources. This prior information might be embodied in knowledgeable individuals, scientific publications, characterization of gene bank holdings, or the findings of molecular biology.

Taken together, economic analysis of genetic enhancement and the search for genetic resources indicate that returns to the discovery and use of crop genetic resources exceed the costs. Many scientists, however, have raised concerns about the continued availability of sufficient genetic resources for future plant breeding efforts. Furthermore, both the scientific and economic literatures agree that the measurement of genetic diversity is complex.

## Factors Influencing Trends in Crop Genetic Diversity

Diverse genetic resources have been a source of large gains in agricultural productivity and, as a result, producer and consumer well-being. Such gains might provide incentives for conservation and efficient use of valuable resources, but these incentives are often muted in the case of genetic resources because returns to their identification and use are not always easily captured by individual farmers, firms, or countries. In fact, the loss of genetic diversity in a species, also called genetic erosion, has been reported in many commercially important crops (National Research Council, 1972; National Research Council 1993; Porceddu et al., 1988).

Genetic diversity is a particular concern because greater genetic uniformity in crops can increase vulnerability to pests and diseases (National Research Council, 1993). Genetic uniformity does not, in and of itself, mean that a particular variety is more vulnerable to pests and diseases or abiotic stresses. In fact, modern varieties often are bred for superior resistance, hence their popularity. Nonetheless, as pests and diseases evolve to overcome host plant resistance, genetic uniformity increases the likelihood that such a mutation eventually will prove harmful to a crop. The evolved pest or disease has a greater crop base that it can successfully attack, which could increase its severity. Instead of a particular disease harming only a small percentage of varieties on limited land, the disease now could affect a greater proportion of a crop's production. For example, genetic uniformity contributed to the spread of the Southern Corn Leaf Blight, which led to a 15-percent reduction in the U.S. corn crop in 1970.

Here, we identify three factors that might contribute to loss of genetic diversity—habitat loss, conversion from landraces to scientifically bred varieties, and genetic uniformity in scientifically bred varieties—and assess how much each factor is operative today. Considerable debate surrounds both the historic and current loss of genetic diversity, due in part to difficulties in defining an appropriate concept of genetic diversity and obtaining accurate measurements. (Formal measures of genetic diversity, as applied both by scientists and by economists, are discussed in the Appendix.) Formal measures of genetic diversity tend to be both wide ranging and data-intensive, and, in most cases, they are not available for long periods (see box, "Measures of Crop Diversity"). As a result, the discussion of trends in genetic diversity is indicative, not precise.

Most of the formal definitions of genetic diversity are applied either at the cross-species level or within a particular species. Within a crop species, these definitions may be related to the number of varieties, the distribution of varieties within a given area, and/or the genetic difference between varieties within a given area or period of time. In the context of crop genetic resources, for example, habitat loss is likely to affect diversity primarily at the cross-species level, where the relevant species are those closely related to the crop of interest. Conversion from landraces to scientifically bred varieties and genetic uniformity in scientifically bred varieties, on the other hand, may affect one or more of these types of indicators *within* a particular crop species.

## Habitat Loss

One factor contributing to a decline in crop genetic diversity has been the loss of wild relatives of cultivated crops (National Research Council, 1993). The loss of wild relatives occurs mainly through habitat conversion for agricultural use. When forest and other wild lands are cleared, plant, animal, and microorganism populations generally fall, reducing the level of genetic diversity. Habitat loss is particularly problematic in developing countries, which often face greater pressures for wild land conversion than do developed countries (Houghton, 1994). Population growth and extensive farming techniques are often cited as factors fostering high rates of land conversion to agriculture. Other influences on land conversion are thought to include poverty, international trade, land degradation, and government policies, particularly where land tenure policies are not clearly defined or enforced (Day-Rubenstein et al., 2000).

Because the full economic values of wild relatives can rarely be captured by landowners, the use of land to preserve habitats for wild relatives remains undervalued compared with alternative uses such as clearing for agricultural or urban use. Thus, habitat conversion occurs in part because the private returns to genetic and other biological diversity are lower than the social returns (Hanemann, 1988). Private returns are important because resources are generally held (whether formally or informally) at the individual or local level. Therefore, many decisions that affect conservation of biodiversity, such as land clearing, are made at these levels. By contrast, many of the benefits of biodiversity conservation accrue at the national or global level. These differing returns contribute to biological resource depletion because conservation of habitat competes with alternative uses of land. Since keeping land in its natural state reduces or eliminates the land's earning capacity for its holders, returns to agricultural production form one opportunity cost of wild land preservation. Also, temporal issues come into play: individuals may place a greater value on current consumption, when weighing the tradeoff between present and future use of resources, than does society as a whole. Together, these factors generate private or individual decisions that differ from those that are socially or globally optimal.

## MEASURES OF GENETIC DIVERSITY

Measures of genetic diversity are very numerous, although there are strong similarities and relationships among many of these measures. At a general level, most involve measures of the *number* of species, the *distribution* of species, and/or the *difference between* species within a given area or period of time. More narrowly, similar concepts might be applied within a crop species, with varieties rather than species becoming the relevant unit of observation.

One reason for the wide variety of measures of genetic diversity is that different people have different reasons for studying or using it. Evolutionary biologists might want to study the process of speciation or the formation of new species, or measure the evolutionary distance between species. Ecologists may be interested in the number and distribution of species within a given habitat. Plant breeders usually focus more closely on diversity within a crop species of interest, although they may also wish to tap diversity within the secondary and tertiary gene pools for that species. (The *secondary gene pool* consists of all biological species that can be crossed with the cultivated species, although these crosses are usually sterile. The *tertiary gene pool* consists of those species that can be crossed with the cultivated species only with difficulty, such as with genetic engineering).

Farmers, particularly those cultivating landraces in noncommercialized agriculture, may be interested in morphological diversity—i.e., diversity in certain physical traits. Because traits are influenced by environmental factors, and because, in many cases, many interacting genes contribute to trait expression, morphological diversity may not be considered to be a "true" measure of genetic diversity. Nonetheless, farmers may make their planting decisions based on such morphological diversity, so it is a potential influence on underlying genetic diversity. Policymakers may focus on preserving genetic diversity as a means to continue crop improvement and guard against the risks of pest or disease epidemics. Economists may wish to study the ways in which the variables important to farmers or policymakers interact with the variables important to plant breeders or ecologists. But no single measure fulfills all desired criteria (Meng et al., 1998).

Also, because certain genetic materials are easy to transport and replicate once collected, it is difficult for countries to capture more than a fraction of the value that flows from their genetic resources. Moreover, markets do not exist for most of the other environmental services provided by biological resources, such as carbon sequestration. Consequently, keeping land in less intensive uses favorable to the *in situ* preservation of genetic resources is often less profitable than more intensive agricultural production to individual countries as well as to individual landowners.

Although many habitat reserves have been established worldwide, wild relatives of agricultural species tend to be included only by accident (FAO, 1996b). Habitat preserves often focus on areas rich in species diversity—usually wildlife species or all plant species—and not on crop species alone. These areas are not necessarily those with the greatest crop genetic diversity.

Much empirical work has focused on the loss of tropical forests, but continued agricultural expansion onto other land is also expected (Day-Rubenstein et al., 2000), although at rates lower than previously projected (Bruinsma, 2003). Compared with the developing world, the developed world has lower rates of agricultural land expansion. For example, the amount of U.S. land used for agricultural production has remained stable since 1945 (ERS, 2002). This does not mean that the same land has been in production. Urban land expansion has displaced some agricultural lands, which have displaced some wild lands. Still, expansion of the agricultural production area has not been a significant factor in U.S. biodiversity loss in recent years.

## Displacement of Landraces by Scientifically Bred Varieties

Crop genetic diversity also declines as landraces are displaced by scientifically developed modern varieties (National Research Council, 1972; Proceddu et al., 1988; Chang, 1994; Kloppenburg, 1988). The ongoing selection process is thought to have narrowed the genetic base of varieties used in agricultural production (Brush, 1992; FAO, 1996b; GAO, 1997; Goodman and Castillo-Gonzalez, 1991). In particular, the spread of high- yielding "Green Revolution" varieties and associated changes in crop management practices beginning in the 1960s is thought to exemplify this transition from landraces to modern varieties (Frankel, 1970; Tilman, 1998). Far less area is planted to landraces worldwide than a century ago. But in many cases, the transition to modern varieties predates the Green Revolution. Improved crop varieties, such as hybrid corn or semi-dwarf wheat or rice, often replaced other varieties that were already the products of scientific crop improvement (see Smale, 1997, for an example). In the broadest sense, alteration and narrowing of crop genetic diversity began with the first domestication of wild plants. For example, the corn plant has been completely dependent on humans for reproduction for thousands of years, because farmer selection has resulted in kernels that can no longer disperse without human intervention.

Farmer choice is a key driving factor behind the replacement of landraces with scientifically bred varieties. When choosing varieties, farmers consider yield potential as well as other production and consumption attributes. Sometimes landraces offer superior yields or resistance to biotic and abiotic stresses, but often they do not. Landraces often provide consumption characteristics traditionally preferred to those of modern varieties (such as maize better suited for tortillas), but even this advantage is not absolute. While maintenance of a diverse set of landrace varieties may prove valuable to current or future plant breeding, individual farmers do not directly capture these benefits, so they have little incentive to account for them when selecting seed for planting. Landraces become extinct through disuse if farmers stop planting and maintaining them, unless stored *ex situ*. Even if many landraces are stored in gene banks, genetic diversity might be lower than if these landraces were planted by farmers, because in the gene bank they are not subject to ongoing evolutionary pressure.

The rate of landrace replacement by scientifically bred varieties differs by crop, world region, and environment. In most industrialized nations, commercialized crops—i.e., crops grown solely for the market, not home consumption—consist almost completely of scientifically bred varieties, although isolated use of landraces may occur.[9] In developing countries, genetic resource specialists often have information about the location of crop landraces and the rate at which they are being replaced by scientifically bred varieties, but published information that is accurate and aggregated is difficult to find.

Some information is available, however, for use of landraces of the three major world cereals, rice, wheat, and corn (maize). In the 1990s, approximately 15 percent of the global area devoted to rice was planted to landraces. Rice landraces are concentrated in southeast Asia, with some also found in the Indian subcontinent (Cabanilla et al., 1999). Use of rice landraces varies by environment and is much lower in the irrigated lowlands than in the more difficult rain-fed lowland and flood-prone and upland environments.

About 10 percent of the developing world's wheat area was planted to land- races in the 1990s. Wheat landraces were concentrated in West Asia and North Africa, with some also found in Ethiopia, China, the Indian subcontinent, and small areas in Latin America. The proportion of wheat area planted to landraces also varied by wheat type and environment. For example, 23 percent of the area planted to durum wheat and 12 percent of the area planted to winter bread wheat was sown to landraces, while only 3 percent of the spring bread wheat area in developing countries was still planted to lan-draces (Heisey et al., 2002).

Unlike wheat and rice, which self-pollinate, corn cross-pollinates, which means that one plant is often fertilized by another. Because of this feature, corn populations are inherently less stable genetically. Therefore, corn landraces may be very diverse genetically. Furthermore, if farmers continue to replant seed (even from hybrids or other scientifically improved corn varieties) rather than buying new seed, the resulting progeny may also be quite genetically diverse. As a result, it is more difficult to define and measure what constitutes a landrace and what is "improved germplasm" for corn than it is for rice or wheat (Morris et al., 1999). That said, it is clear that a far higher percentage of the developing world's corn area (just under 40 percent) is planted to landraces than is the case for either wheat or rice. If developing countries that produce primarily temperate corn or countries that market "commercialized" corn[10] are excluded, nearly 60 percent of the developing world's corn area is planted to landraces (Morris, 2002). As with the other cereals, corn's wild relatives tend to concentrate in their zone of origin (in the case of corn, in Mexico and Central America), and landraces are most diverse in this zone. Nonetheless, corn landraces are found in many parts of the developing world.

## Genetic Uniformity in Scientifically Bred Varieties

In situations where most or all landraces have been replaced by scientifically bred varieties, crop genetic diversity may also decline with (1) reductions in total numbers of varieties, (2) concentration of area planted in a few favored varieties, or (3) reductions in the genetic distance between these varieties. The National Research Council (1993) concluded that the genetic vulnerability of U.S. wheat and corn has become less of a problem since 1970, in part because of efforts to breed in greater diversity. However, the Council also determined that genetic uniformity of rice, beans, and many minor crops is still a concern.

Information for other countries is not readily available. Relatively little attention has been paid to genetic uniformity of scientifically bred varieties in developing countries, perhaps because there more focus has been placed on habitat conversion and displacement of landraces. One major study, however, analyzed trends in modern spring bread wheats planted in the developing world, both in the genetic diversity of varieties released and varieties planted in farmers' fields (Smale et al., 2001). This study was representative of over 50 million hectares of wheat planted in the developing world. Both pedigree analysis and molecular analysis suggested that the genetic diversity of these modern wheat varieties had increased, not decreased, over the past 30 years. Trends in genetic diversity for other crops in developing countries, however, as well as for crops in industrialized nations outside the United States, would likely vary by crop and region.

Whatever the trends in genetic diversity, the genetic uniformity of many crops has raised concerns that crop yields and production will become more variable from season to season (Swanson, 1996). As with other drivers of genetic erosion, individual farmers have limited incentives to consider the wider potential consequences of genetic uniformity, and, when choosing which varieties to plant, may perceive the benefits of uniform varieties to be greater. Farmers may be willing to accept the risk of greater variability if they expect to receive higher average yields.

Thus far, despite concerns about genetic uniformity, yields for many major crops have been relatively stable. An important reason may be that temporal diversity has replaced spatial diversity (Duvick, 1984). Although there may be greater spatial uniformity of crops planted at any given time today (compared with 100 years ago), modern plant breeding provides a steady release of new varieties with new traits for pest or disease resistance over time.

The ability of plant breeders to keep ahead of evolving pests and diseases through temporal diversity depends directly on the quality and accessibility of germplasm collections in public gene banks and in private breeders' collections. Because many of the benefits of raw germplasm cannot be appropriated, private breeders rely on the public sector to collect, characterize and perform pre-breeding enhancement of genetic materials to make them accessible for private use (Duvick, 1991).

## CONSERVATION OF PLANT GENETIC RESOURCES

In this section we examine two basic strategies for conserving genetic resources, three principal tools policymakers can use to support these strategies, and several multilateral agreements by which countries currently seek to coordinate international use of these tools. Decisions about these alternatives may affect U.S. access to genetic resources that are currently held outside the United States (and vice versa).

**Table 1. Advantages and disadvantages of *ex situ* versus *in situ* conservation**

| ***Ex situ* conservation** | | ***In situ* conservation** | |
|---|---|---|---|
| **Advantages** | **Disadvantages** | **Advantages** | **Disadvantages** |
| Costs generally centralized | Certain types of germplasm not readily conserved | Genetic resour-cees used to pro-duce valuable product | Costs borne by farmers (for landraces) |
| Can preserve large amounts of diverse germplasm | Regeneration can be costly, time-consuming | Evolutionary processes continue | May reduce on-farm productivity |
| Germplasm can be readily accessed by more breeders | Potential for genetic "drift" can reduce integrity of collection | May better meet the needs of certain farmers | Requires land |
| High-security storage imper-vious to most natural disasters. | In practice, many collections lack the resources nee-ded to organize, document, and maintain their samples. | More efficient for some germ-plasm, e.g., animals, or crops that reproduce vegetatively. | Farmer selections may not preserve targeted diversity |
| | | Existing wild relatives can be preserved with-out collection | Loss of wild relatives when land use changes |

Note: The pointer locations indicate general regions where crops are believed to have first been domesticated. In some cases, the center of origin is uncertain. Other geographic regions also harbor important genetic diversity for these crops.

Source: This map was developed by the General Accounting Office using data provided by the National Plant Germplasm System's Plant Exchange Office.

Figure 3. Centers of origin of selected crops

## Basic Conservation Strategies

At the most basic level, genetic resources can be conserved either *in situ* (in their natural setting) or *ex situ* (outside their natural setting). *In situ* is the dominant method of conserving natural ecosystems. Crop genetic resources are commonly held *ex situ*, but they can also be held *in situ*—as wild relatives of cultivated varieties on wild land and as cultivated varieties in farmers' fields. Among the decisions policymakers face is the appropriate balance between *in situ* and *ex situ* conservation efforts. Each has its own benefits and drawbacks; the two are perhaps better viewed as complementary rather than as substitutes (table 1).

### *In situ conservation*

Species preserved *in situ* remain in their natural habitat. Most of the world's genetic diversity is found *in situ*. For agriculturally important species, the greatest diversity in landraces and in wild relatives is typically found near where they were first domesticated. Early in the twentieth century, Russian botanist N. I. Vavilov defined "centers of origin" for most crops. These included Mexico and Central America (for corn, or maize as it is known in the rest of the world, and upland cotton); China (for soybeans); and West Asia (for wheat and alfalfa).

Since Vavilov's time, ideas about centers of origin have been refined. Some crops, such as sorghum, sugarcane, and peanuts, were probably domesticated over very broad areas rather than in a well defined center (Harlan, 1971, 1992). Furthermore, useful landraces of some crops have been found in parts of the world other than those in which they were originally domesticated. For example, wheat landraces found in the pedigrees of many modern wheat varieties have come from every continent except Antarctica (Smale and McBride, 1996).[11] Still, *in situ* preservation efforts, as well as germplasm collection activities for *ex situ* conservation, are often focused most closely in and around centers of origin (figure 3).

Because *in situ* conservation of agricultural genetic resources is carried out within the ecosystems of farmers' fields or wild lands, species continue to evolve with changing environmental conditions. *In situ* conservation thus can provide valuable knowledge about a species' development and evolutionary processes, as well as how species interact. By allowing genetic resources to act as part of larger ecosystems, *in situ* conservation may also provide indirect ecological benefits, such as hosting diverse pollinators. However, since restrictions on land use may be necessary, *in situ* conservation can be costly. To conserve agricultural genetic diversity *in situ,* for example, a farmer may have to forgo the opportunity to grow a higher yielding (and more profitable) variety. Or, in the case of wild *in situ* resources, the land may need to be set aside from agricultural production or other production-related uses completely. This suggests one important constraint on *in situ* conservation that has been addressed in our discussion of habitat loss—the divergence between the social and private returns to conserving genetic diversity.

### *Ex situ conservation*

The *ex situ* method removes genetic material from its environment for longterm conservation (table 1). Botanical gardens and gene banks are examples of *ex situ* conservation strategies. Certain methods of *ex situ* conservation can be used to store large amounts of genetic material at relatively low cost, certainly in terms of land needed, compared with *in situ* strategies. The world's gene banks presently hold more than four million accessions, or specific samples of crop varieties. It is estimated that samples of many of the world's cereal landraces are now held in gene banks (Plucknett et al., 1987). Although very few important crop species originated in what is now the United States, the U.S. national gene bank system (the National Plant Germplasm System, or NPGS) is today one of the largest *ex situ* collections in the world. *Ex situ* conservation also is appealing because it allows plant breeders easier access to genetic resources than is provided by *in situ* conservation.

However, crop genetic resources first must be collected, and samples of only a small fraction of the world's plant genetic resources have been collected thus far. Stored plant materials must be kept under controlled conditions, and periodically regenerated (planted and grown) in order to maintain seed viability. Not all kinds of plant genetic resources are easily conserved *ex situ*. Some lose their varietal identity when stored as seed. These plants may need to be kept as living plants, a more costly process that requires additional land and labor. And gene banks in politically unstable areas may be in danger of losing valuable genetic material. Even in stable locations, the resources necessary to maintain or improve plant gene banks are not always forthcoming because of competing demands for public resources (GAO, 1997).

## Policy Tools to Promote Genetic Resource Conservation

Three major types of policy tools are available to support conservation of genetic resources: (1) public investment in *in situ* and *ex situ* conservation; (2) stronger intellectual property rights over genetic inventions, particularly in developing countries; and (3) material transfer agreements.

### *Public funding of ex situ and in situ conservation*

Funding conservation is the most direct method of preserving crop genetic resources. Past efforts have convinced plant breeders that the current germplasm stock, if properly maintained, is adequate to maintain steady yield growth over the next 20 to 50 years (Shands, 1994; Sperling, 1994; Siebeck, 1994). There is growing concern, however, that this may not be sustainable in the long term at current funding levels (Keystone Center, 1991; NRC, 1993; OTA, 1987; FAO, 1996b). Studies of gene banks worldwide (FAO, 1996a), the U.S. National Plant Germplasm System (GAO, 1997), and the Vavilov Institute collection in the former Soviet Union (Zohrabian, 1995) conclude that most gene banks lack sufficient funds, facilities, and staff to maintain their germplasm collections.[12] Funding problems arise, in part, because individual nations do not capture the full benefits of investments in genetic resource conservation. While multilateral funding of international crop research facilities has been used to alleviate this problem, free rider problems suggest that funding for international facilities will remain less than optimal.

The UN Food and Agriculture Organization (FAO) reported on the most pervasive problems facing gene banks worldwide (FAO, 1996a). First, since 1970, more emphasis has been placed on collecting materials, than maintaining accessions, and most gene banks lack adequate long term storage facilities. Even accessions in suitable long term storage cannot be maintained indefinitely; collected material must be grown out or "regenerated" periodically. Many gene banks lack the funds, facilities, or staff to carry out needed regenerations. Second, while gene bank coverage of elite and landrace varieties of major cereal crops is believed to be fairly complete, coverage of many "minor" crops (such as root crops, fruits, and vegetables) and wild relatives remains spotty. Third, only a small fraction of accessions has been characterized. This lack of information about what actually resides in these collections constrains breeders from using new genetic materials (NRC, 1993) and makes it difficult to identify gaps in collections. Fourth, many countries have reported that funding has been unstable and uncertain year to year, hampering investment and planning decisions. The FAO (1996c) concluded that "without prompt and significant intervention, much of the stored genetic diversity of food and agricultural crops in the world-as well as the large public investment made in assembling the collections-will be lost forever."[13]

The same public goods problem that inhibits optimal international investment in *ex situ* conservation of genetic resources-the inability of conserving nations to capture all the benefits from that conservation-also hinders optimal investment in *in situ* conservation. Moreover, *in situ* conservation is subject to several additional constraints. First, uncertainty surrounding the likely magnitudes of the benefits of in situ conservation is probably larger than it is for *ex situ* conservation. Second, the number of economic agents and levels involved in any *in situ* conservation effort (including landowners and/or individuals with rights to use the land) is likely to be considerably larger than for *ex situ* programs, making coordination of *in situ* programs more difficult.

*In situ* conservation of wild relatives and landraces require different strategies. Establishing habitat reserves could protect wild relatives. Turkey, for example, has received multilateral funding for an *in situ* pilot project to conserve wild relatives of wheat and barley (FAO, 1996b). For landraces, if farmers have private incentives to maintain local varieties, policy interventions for *in situ* conservation may be unnecessary. In areas where displacement of local varieties is more likely, access to modern varieties need not be completely prohibited. A less costly alternative might be to establish some type of conservation easement, paying local farmers the difference between returns to modern and local varieties if they grow a diverse set of varieties on part of their plots (Christensen, 1987). Yet another approach could be to purchase limited amounts of landrace seed from producers in regions with diversity

Most experts agree that *in situ* and *ex situ* conservation strategies are complementary, however the best allocation of resources is subject to debate. Plant breeders are concerned that increased investment in *in situ* conservation will compromise gene bank maintenance. Lack of data on the relative costs and benefits of *in situ* and *ex situ* conservation increases the difficulty of allocating funds across activities. Moreover, donor institutions, particularly at the national level, face competing needs, some of which offer more direct and immediate benefits.

### *Intellectual property rights*

Adoption of stronger intellectual property rights (IPR) regimes has been one of the most commonly proposed methods to enhance genetic resource conservation internationally. Proponents argue that stronger IPR will allow the holders of genetic resources to reap the rewards from commercializing these resources and thus align private incentives more closely with public incentives for genetic resource conservation.

Historically, the set of IPR used for genetic resources internationally focused on the products of formal plant breeding programs rather than wild relatives and landraces. Even while varieties developed by breeders were protected by formal "plant breeders' rights", wild relatives and landraces continued to be considered a public good. For decades, many plant breeders have freely exchanged "raw" germplasm (Kronstad, 1996; Heisey et al., 2001).[14] National plant breeding programs and international agricultural research centers freely provide such unshielded genetic materials not only to other public breeding institutions but also to private breeders (many of them in developed countries) who may then use those materials to develop new commercial crop varieties for sale (Day, 1997).[15]

This asymmetry has proven controversial. Many developing countries and nongovernmental organizations (NGOs) make the case for "farmers' rights," arguing that farmers in developing countries have selected and saved landraces for thousands of years, making an essential contribution to plant breeding and crop variety development (Mooney, 1979, 1983; Brush, 1992). It is unfair, they argue, that private breeders have free use of wild relatives and landraces but require payment for elite varieties based, in part, on germplasm that originated in developing countries. Others counter that the exchange of genetic material for plant breeding has been beneficial to developed and developing countries alike, although they disagree about whether foregone earnings from sales of raw genetic material by lower-income countries are compensated for by other benefits, such as unrestricted access to public germplasm and lower food prices for consumers (Shands and Stoner, 1997; Fowler, 1991).

Proponents of stronger IPR regimes argue that, generally speaking, they encourage commercialization of genetic resources, thus enhancing the incentives for conservation, both *in situ* and *ex situ*. They also maintain that greater IPR stimulate private sector research, relieve public budgetary constraints, and increase national incentives for germplasm conservation (Barton and Siebeck, 1991). Critics counter that stronger IPR would do little to increase innovation or maintain crop genetic diversity, arguing that private incentives favor specialization and product uniformity rather than diversity in the production of new seed varieties (Mooney, 1979, 1983; Acharya, 1991; Reid, 1992; Brush, 1994).

These arguments raise two empirical questions. First, what impact would stronger IPR protection have on germplasm use and exchange? A survey of 84 private plant breeding firms by Pray et al. (1993) assessed the impacts of a 1985 decision by the U.S. Supreme Court that strengthened genetic resource IPR (for modern varieties) by allowing plant breeders to acquire utility patents for new varieties.[16] More than a third of the firms felt that utility patents limited germplasm exchange both between private firms and between the public sector and private firms. Six of 84 firms reported that they had increased their research expenditures because of the availability of utility patent protection. Most reported that utility patents increased profitability. Rejesus et al. (1996) surveyed wheat breeders internationally, and reported that respondents believed that stronger international IPR for plant varieties would reduce germplasm exchange between developed and developing countries, reduce exchange between developing countries and reduce the use of foreign landraces. Pray (1990) noted that stronger IPR in developing countries would entail significant enforcement costs and other transaction costs. The effect of IPR targeted toward land races and wild relatives remains unknown.

Second, what are the implications of stronger IPR and increased private R&D for the diversity of new varieties developed? Some evidence suggests that the diversity of major crops has not declined in the United States as increasingly strong IPR protections have been enacted over the last 30 years, and diversity may have actually increased for some crops (Duvick, 1984; NRC, 1993; Smale and McBride, 1996; Falck Zepeda and Traxler, 1997; Pray and Knudson, 1994; Knudson, 1998). But the role of IPR is confounded by other efforts to increase crop genetic diversity (see NRC, 1993, pp. 67-81, for discussion on the impacts of its 1972 report "Genetic Vulnerability of Major Crops").

### *Material Transfer Agreements*

Material transfer agreements (MTAs) are legal instruments initially used as a means for transferring biological materials between entities, including public institutions, private companies, and countries. Initially, used for research only, MTAs may be bilateral agreements or may follow a standard template (such agreements are often used by public entities). The provider retains commercial rights to the material. MTAs have become a common instrument to outline the terms for sharing genetic resources and, sometimes, the gains from new product development. MTAs may include provisions for intellectual property rights, such as what, if any, IPR may be sought for the transferred material or inventions based on that material. However, not all MTAs address IPR and even if they do, IPR usually are just one element of the agreement.

Interest in using MTAs as an incentive to preserve germplasm stems from the idea that benefit sharing can reward suppliers of genetic resources (Barton and Christensen, 1988; Blum, 1993; Christensen, 1987; Simpson and Sedjo, 1992; U.S. Department of State, 1994; WRI, 1993). The benefits to be shared may include funds, materials, training, technology, or intellectual property rights (through provisions concerning their allocation).

The potential for benefit sharing MTAs to affect crop genetic resource conservation is unclear. Plant breeders of major crops use germplasm mainly from their own working collections, or acquire it from other breeders, botanists, or geneticists. Typically, this germplasm has already been enhanced and adapted for plant breeding purposes. While exotic germplasm may provide especially useful traits for disease or pest resistance, such germplasm is only one source of the many genes used in an individual variety. Statistics suggest that, for many commercially important crops, only a small percentage of the genes in released varieties are from newly incorporated exotic germplasm (Cox et al., 1988; Goodman and Castillo-Gonzalez, 1991). The expected value of such exotic germplasm is generally small, though on occasion benefits may be larger (Wilkes, 1991). When breeders do require genetic traits unavailable from their conventional sources, gene banks such as the Future Harvest

Centers or the NPGS traditionally have had a vast, free supply of germplasm. To date, this germplasm has been provided freely to users, and not subject to MTAs that require benefit sharing. The use of MTAs to market germplasm from some developing countries may also be hindered by a lack of technical expertise. Breeders often require documentation of valuable genetic traits and the ease by which they can be transferred to commercial seed stock. Even if a country has rare and useful germplasm, breeders may remain unaware of its value or existence (Shands, 1994).

To date, the use of MTAs for crop genetic resources has not generated large financial gains for developing countries. In this respect, raw genetic resources, though lacking a well-developed market, are similar to primary export commodities such as timber or coffee. Much of the value added to commercial seed varieties comes from the laborious and time-consuming process of incorporating raw genetic material into elite crop varieties.

## Multilateral Agreements Affecting Plant Genetic Resources

Because of the widespread geographic origins and current use of crop genetic resources and the public goods nature of their conservation, the three principal policy tools for conserving genetic resources involve considerable international overlap. A series of multilateral agreements embody the international coordination needed to preserve genetic resources, as well as the lingering debate over property rights for genetic resources.

### *U.N. Convention on Biological Diversity*

The 1993 U.N. Convention on Biological Diversity (CBD) was designed to promote the conservation and sustainable use of biological diversity and to encourage the equitable sharing of resulting benefits. Language in the Convention relating to property rights over genetic materials, biological inventions, technology transfer, and benefit sharing was drafted more with pharmaceutical and industrial development in mind than seed variety development, though subsequent meetings to implement the Convention focused on agricultural biodiversity. On December 29, 1993, the CBD came into force for ratifying and acceding parties (which numbered 188 as of February 15, 2005).[17] Provisions of the Convention have direct implications for the collection, preservation, and exchange of genetic resources. The CBD states that countries have sovereign rights to their indigenous genetic resources, which institutionalizes the change from the practice of freely collecting and sharing of resources. Most countries have interpreted the CBD to allow countries to require payments or transfer of technology in exchange for access to germplasm. The Convention also included a provision for a biosafety protocol to regulate the international movement of the products of biotechnology. Adopted in January 2000, the "Cartagena Protocol" addresses only living modified organisms (LMOs), and makes a distinction between genetically modified organisms as seed and genetically modified organisms intended for food or feed (the assumption being that the latter will not be released into the environment). According to the protocol, LMOs (which include genetically modified seed) are subject to "Advanced Informed Agreement" procedures. Thus, implementation of the protocol has more impact on LMOs that are transferred as seed, or as germplasm for use in genebank system, than on food or feed.

Other agreements play a role. The World Trade Organization (WTO) agreements, which are negotiated, signed, and ratified by the bulk of the world's trading nations, are enforceable through the WTO's ability to levy sanctions. Therefore, countries have strong incentives for the CBD to be consistent with the Trade Related International Property (TRIPS) provisions and the WTO. The International Union for the Protection of New Varieties of Plants (UPOV) is another element affecting the exchange of genetic resources. UPOV-consistent IPR are the leading form of formal varietal protection globally (UPOV protection allows exemptions for breeding and research purposes). After the CBD came into force, the U.S. Department of State (1994) noted that the Convention could not be used to overrule existing intellectual property law, including TRIPS and UPOV. Therefore, both are likely to continue influencing implementation of the CBD.

### *The International Treaty on Plant Genetic Resources for Food and Agriculture*

To address issues left unresolved by the CBD, the International Treaty on Plant Genetic Resources for Food and Agriculture was developed with the intention of (1) mandating conservation of plant genetic resources, (2) ensuring equitable sharing of the benefits created by using these resources, and (3) establishing a multilateral system to facilitate access. The International Treaty entered into force in June 2004 (the U.S. has signed, but not yet ratified the treaty). Sixty-six countries are parties to the treaty. The treaty is to govern international exchange of germplasm and will cover 35 crops, including major cereals like rice, wheat, and maize, but excluding soybean and peanut and other important crops.

IPR have been a major source of debate in interpreting the treaty, particularly the patenting of materials discovered in public gene banks. The treaty states that "Recipients shall not claim any intellectual property or other rights that limit the facilitated access to the plant genetic resources for food and agriculture, [or their genetic parts or components,] [in the form] received from the Multilateral System." Interpretations of this clause abound, particularly with respect to whether the patenting of isolated compounds, such as genes, will be permitted.

The treaty is vague on a number of points. Disagreements remain about the implementation of benefit sharing and the development of a standard Material Transfer Agreement (MTA). The standard MTA is intended to establish the terms of access to plant genetic resources, and all germplasm exchanges under the new multilateral system will be governed by this standard MTA (rather than the bilateral approach suggested by the CBD). The benefits arising from commercial use of germplasm accessed under the multilateral system are to be shared through four mechanisms: (1) exchange of information, (2) access to and transfer of technology, (3) capacity building, and (4) sharing of monetary and other benefits of commercialization. A yetto-be established portion of monetary benefits from commercial products are to flow, through a trust account managed by the Governing Body of the Treaty, primarily to farmers who conserve genetic resources, especially those in developing and transitional economies.[18] Because benefits will be shared according to conservation practices and income, rather than contributions to the multilateral system, the incentives for conserving genetic resources are likely to be less direct than originally envisioned. More broadly, the means and particulars of financing conservation activities also have not been specified.

## Financing International Conservation of Genetic Resources

Given the public good characteristics of crop genetic resources, financing their conservation remains a challenge. Resources available under current and immediately foreseeable policies may be insufficient to conserve the resources agriculture will need. Though MTAs and the expansion of IPR are intended to be self supporting conservation policies, proposals to intensify *in situ* and *ex situ* conservation and to transfer technology and expertise would require additional public funds. Various efforts have been made to estimate actual amounts needed to finance gene banks, *in situ* preservation, and technology transfer. The Keystone International Dialogue (Keystone Center, 1990, 1991) recommended a fund of $300 million annually to support global and national efforts to conserve plant genetic resources. The U.S. National Research Council (1993) recommended that $240 million would be needed annually for maintaining worldwide base collections in addition to evaluation and documentation programs. The FAO (1997) estimated low (A), medium (B), and high (C) funding options ranging from $150 million to $248 million to $455 million annually, averaged over more than ten years. The FAO figures include only costs that would be borne by the international community and do not include domestic program funding. The report considered Option A "basic or rudimentary" while Option B was "consistent with known and documented needs and realistic absorption and implementation capacity of countries" (FAO, 1997).

Grounded in the FAO's Global Plan of Action for genetic resources is a relatively new organization focused more directly on *ex situ* genetic resource preservation. The Global Crop Diversity Trust is an international organization whose establishment has involved a partnership with the FAO and the 16 Future Harvest Centers of the Consultative Group on International Agricultural Research (CGIAR). The Trust aims to match the long-term nature of conservation needs with permanent, sustainable funding by creating an endowment that will perennially fund crop diversity collections around the world. The endowment is intended to facilitate the perpetual conservation of eligible collections that meet agreed standards of management. The Trust will serve as an element of the funding strategy to be implemented under the International Treaty described above.

The Global Crop Diversity Trust hopes to raise a minimum of $260 million from corporations, trusts, foundations, and governments as a permanent endowment for genetic resources. That figure is based on a study carried out by the International Food Policy Research Institute and the University of California, Berkeley, which provided best estimates of the annual funds needed to support the core services provided by the Future Harvest genebanks and the level of endowment needed to provide for the collections in perpetuity (Koo et al., 2002). (The annual costs were estimated to be $5.7 million, and the needed endowment was estimated to be $150 million). The Trust has approximately $45 million in commitments and $70 million under discussion to date (Global Crop Diversity Trust, 2005).

Some researchers have looked at methods beyond multilateral donor systems to fund conservation of genetic resources. Proposals have included a tax on seed sales to provide funds for conservation (Barton and Christensen, 1988). Barton and Christensen suggested either a "straight" sales tax on seed revenues or a system of royalty calculation similar to that used by record companies, with proceeds to be distributed among international, national, and private conservation programs to fund *in situ* and *ex situ* preservation. There are concerns that a royalty based system of direct payments may limit the exchange of genetic resources. Also, if royalty payments in the strict sense are used (i.e., payment upon use in a released variety), returns probably will be limited (Charles, 2001). Proposals to fund germplasm through sales taxes and user fees have been opposed by private seed companies. Even if the proposals were to overcome this opposition, formal seed sales are much less prevalent in self-pollinated crops and in some crops grown in developing countries. Thus, certain crops would not benefit as significantly from this approach.

Another proposal has been to tax agricultural commodities generally (Swaminathan, 1996). This proposal raises questions about the distributional implications (between regions and social classes within regions) of taxing seeds or all agricultural commodities. Because poor families generally spend more on food as a portion of the household budgets, such taxes may be regressive (though to raise equal revenues, the tax rate for a general commodity tax on agricultural, forest, and fish products would need to be only a small fraction of the tax rate for seeds). Another option lies with agricultural producer groups in developed countries (such as Australia, New Zealand, and the United States), many of which fund commodity specific research and market promotion through voluntary checkoff systems that act as a commodity tax. However, while national producer groups may be persuaded to help support domestic gene banks and germplasm characterization, they may be less willing to allocate checkoff funds to an internationally administered fund. As with other aspects of genetic resource use and conservation, private interests do not necessarily coincide with broader public goods.

## Appendix: Measuring Crop Genetic Diversity

Evolutionary or ecological measures of genetic diversity focus particularly on genetic similarity or difference between different species. These kinds of comparisons might also be useful in the study of crop genetic diversity, particularly if a given crop is analyzed in the context of its wild relatives.[19] However, most studies of crop genetic diversity are based on the similarity or difference between different crop populations within the same crop species. Most commonly, named varieties are the crop populations in question, although two distinct varieties may in fact be very similar genetically (Meng *et al.*, 1998). For the rest of this discussion, we will usually assume diversity is being measured within a particular crop species.

## Spatial Diversity Measures

Spatial diversity—diversity within a given geographical area—may be "the most commonly recognized concept of diversity" (Meng et al., 1998). Two concepts are often used in spatial measures of genetic diversity. "Richness" refers to a simple count measure, for example of the number of varieties of a particular crop species planted in a given area. "Abundance" is a measure of the evenness of the spatial distribution of elements of the set being considered (Magurran, 1991). For example, suppose the same ten crop varieties are planted in two identical regions. In one region, each variety is planted on one-tenth the area, but in the other region one variety is planted on 91 percent of the area and the other nine varieties occupy one percent each. By a simple count measure (such as richness), the two regions are equally diverse, but introducing abundance would suggest the first region is more diverse than the second. This, along with the fact that named varieties may be very similar genetically, is why simply counting numbers of varieties is likely to be an inadequate measure of crop genetic diversity. Simple diversity indices that reflect varietal distribution (thus partially capturing the concepts of richness and abundance), include the proportion of area planted to the most popular variety or given number of varieties (equivalent to concentration measures used in the industrial organization literature.) A related index is the number of varieties covering a given percentage of total crop area (Widawsky, 1996). Another measure taken from the industrial organization literature is the Herfindahl index, which illustrates the degree of concentration among varieties (Pardey et al., 1996). The Simpson index (one minus the Herfindahl index) and the Shannon-Wiener index, taken from information theory, are often applied in ecological studies of diversity (Magurran, 1991).

## Measures of Relationships between Varieties

Other indices of genetic diversity are built up from measures of "genetic distance," i.e., the degree to which varieties or species differ genetically (Nei,

1972; Cavalli-Sforza and Edwards, 1967; Reynolds, Weir, and Cockerham, 1983; Gregorius, 1978). To a certain extent such measures address the problem raised by simply counting named varieties that may be very similar genetically. Genetic distance indices can be calculated based on observations of different crop characteristics, including morphological indicators such as plant height, grain weight, and so on. As indicated, morphological indicators have the advantage that they may be closely linked to the traits on which farmers base their decisions, but the disadvantage that they are often influenced by environment and multiple genes, and therefore not reflective of genetic distance at the chemical (enzyme) or molecular (DNA) level. Genetic distance indices have perhaps most commonly been applied to this biochemical information. The use of biochemical and molecular markers requires systematic physical sampling as well as laboratory time and materials, and as a result can be quite costly (Meng et al., 1998).[20] An alternative approach to measuring genetic distance between varieties, at least for scientifically-bred crops with documented pedigrees, is based on comparison of the heritage of pairs of varieties.[21]

## Building Diversity Indices

Genetic distance indices measure differences between different crop varieties or species, but they themselves do not measure overall genetic diversity. Weitzman (1992; 1993) describes a diversity index calculated as the total length of the branches of a taxonomic tree. Such a tree could be calculated using morphological, genealogical (i.e. pedigree), or genetic distance data. Solow, Polasky, and Broadus (1993) also incorporate the size of the set (e.g., number of crop varieties) as well as genetic distance into genetic diversity indices. Both these tree-based measures, and other measures based on matrices of similarity coefficients, permit weighting to reflect the distribution of crop varieties (Souza et al., 1994; Meng et al., 1998).

## Measures of Plant Breeding Activity Using Genetic Resources

A number of other measures have been applied to the study of genetic resources, but they usually refer to aspects of a scientific plant breeding program, or the development of such a program from initial crosses involving landraces, rather than to direct measures of genetic diversity. These include numbers and origin of landraces in the ancestry of the varieties being studied, or the number of breeding generations since the initial cross (Gollin and Evenson, 1990); numbers of distinct parental combinations and numbers of unique landrace ancestors per pedigree (Smale and McBride, 1996; Hartell, 1996; Smale et al., 1998); or coefficient of parentage (COP) based measures (Pardey et al., 1996). Note that all of these pedigree-based measures are less useful in a crop, such as corn, that may not always follow a strict pedigree breeding system, or in crops for which pedigrees are partially or completely private for proprietary reasons.

## Temporal Diversity

Duvick (1984) observed that in a number of scientifically-bred crops, *temporal* diversity (or diversity through time) has replaced spatial diversity as one means of maintaining or even raising resistance or tolerance to pests and diseases. Temporal diversity depends on maintaining breeding effort by humans. Meng et al. (1998) closely identify temporal diversity with "the rate of change or turnover of [planted] varieties" as defined, for example, by Brennan (1984) and Brennan and Byerlee (1991). Other things being equal, faster varietal turnover might be expected to be associated with increased temporal genetic diversity, but like pedigree-based measures, varietal turnover is more a measure of the output of a plant breeding program than of genetic diversity *per se.* Newly released varieties might be genetically somewhat dissimilar to older varieties, or they might be very closely related genetically. Time-series of *spatial* diversity measures could provide useful information about *temporal* change in diversity, but such a series would not strictly measure "temporal diversity." More formal assessment of temporal genetic diversity could be made by statistically testing differences between genetic distance measures over temporal samples (See Souza et al., 1994 and Tessier and Bernatchez, 1999).

# References

Acharya, R. (1991). "Patenting of Biotechnology, GATT and the Erosion of the World's Biodiversity," *Journal of World Trade,* Vol. 25,71-87.

Ahearn, M., Yee, J., Ball, E. & Nehring, R. (1998). *Agricultural Productivity in the United States.* AIB-740. Economic Research Service, United States Department of Agriculture.

Alston, J. M., Norton, G. W. & Pardey, P. G. (1995). *Science Under Scarcity: Principles and Practice for Agricultural Research Evaluation and Priority Setting*. Ithaca, New York: Cornell University Press.

Attwood, T. K. (2000). "The Babel of Bioinformatics." *Science, 290*, 471-473.

Barbier, E. B., Brown, G., Dalmozzone, S., Folke, C., Gadgil, M., Hanley, N., Holling, C. S., Lesser, W. H., Maler, K. G., Mason, P., Panayotou, T., Perrings, C., Turner, R. K. & Wells, M. (1995). "The Economic Value of Biodiversity," *Global Biodiversity Assessment,* V. H., Heywood (ed.). Cambridge and New York: Cambridge University Press for the United Nations Environment Program.

Barton, J. H. & Christensen, E. (1988). "Diversity Compensation Systems: Ways to Compensate Developing Nations for Providing Genetic Material," *Seeds and Sovereignty: The Use and Control of Plant Genetic Resources*, J. R. Kloppenburg (ed.). Durham, NC: Duke University Press.

Barton, J. & Siebeck, W. (1991). *Intellectual Property Issues for the International Agricultural Research Centers: What are the Options?* Report prepared for the CGIAR Secretariat, Washington, DC.

Blum, E. (1993). "Making Biodiversity Conservation Profitable: A Case Study of the Merck/INbio Agreement," *Environment*, Vol. *35*, No. 4.

Brennan, J., Singh, I. P. & Lewin, L. G. (1997). "Identifying International Rice Research Spillovers in New South Wales," *Agricultural Economics*, Vol. *17*.

Brennan, J. P. (1984). "Measuring the Contribution of New Varieties to Increasing Wheat Yields," *Review of Marketing and Economics*, Vol. 52.

Brennan, J. P. & Byerlee, D. (1991). "The Rate of Crop Varietal Replacement on Farms: Measures and Empirical Results for Wheat," *Plant Varieties and Seeds*, Vol. *4*.

Brown, G. (1987). "Valuation of Genetic Resources" *The Preservation and Valuation of Biological Resources*, G. Orians, G. Brown, W. Kunin, J. Swierzbinski (eds.). Seattle. WA: University of Washington Press.

Brown, G. & Swierzbinski, J. (1985). "Endangered Species, Genetic Capital, and Cost-Reducing R&D," *Economics of Ecosystem Management,* D. Hall, N. Myers, and N. Margaris (eds.).

Bruinsma, J. (ed.) 2003. *World Agriculture: Towards 2015/2030. An FAO Perspective*. London: Earthscan Publications, Ltd.

Brush, S. B. (1992). "Farmers' Rights and Genetic Conservation in Traditional Farming Systems," *World Development*, Vol. *20*, No. 11.

Brush, S. B. (1994). *Providing Farmers'Rights through In Situ Conservation of Crop Genetic Resources*. Report to the FAO Commission on Plant Genetic Resources. Rome: U.N. Food and Agriculture Organization.

Byerlee, D. & Traxler, G. (1995). "National and International Wheat Improvement Research in the Post Green Revolution Period: Evolution and Impacts." *American Journal of Agricultural Economics*, Vol. *77*.

Cabanilla, V. L., Hossain, M. & Khush, G. S. (1999). "*Diffusion of Breeding Materials and Genetic Composition of Improved Rices in South and Southeast Asia 1965-98*." Paper presented at the Impacts Assessment and Evaluation Group (IAEG) Germplasm Research Impact Study Workshop, Nashville, TN, Aug. 7-8.

Cavalli-Sforza, L. L. & Edwards, A. W. F. (1967). "Phylogenetic Analysis: Models and Estimation Procedures.." *American Journal of Human Genetics*, Vol. *19*.

Chang, T. T. (1994). "The biodiversity crisis in Asia crop production and remedial measures," *Biodiversity and Terrestrial Ecosystems*.C.I. Peng and C.H. Chou (eds.), Institute of Botany, Academia Sinica Monograph Series No. 14. Taipei: Academia Sinica.

Charles, D. (2001). "Seeds of Discontent," *Science*, Vol. *294* (October).

Christensen, E. (1987). "Genetic Ark: A Proposal to Preserve Genetic Diversity for Future Generations," *Stanford Law Review*, Vol. *40*.

Cooper, J. C. (1998). "The Economics of Public Investment in Agro-biodiversity Conservation," *gricultural Values of Plant Genetic Resources*. In R.E. Evenson, D. Gollin, and V. Santaniello (eds.), Wallingford, UK: CABI Publishing.

Cox, T., Murphy, J. & Goodman, M. (1988). "The Contribution of Exotic Germplasm to American Agriculture," *Seeds and Sovereignty: The Use and Control of Plant Genetic Resources*. J.R. Kloppenburg (ed.). Durham, NC: Duke University Press.

Cox, T. S., Kiang, Y. T., Gorman, M. B. & Rodgers, D. M. (1985). "Relationship Between Coefficient of Parentage and Genetic Similarity Indices in the Soybean," *Crop Science*, Vol. *25*.

Day, P. R. (1997). "Biodiversity and the Equitable Use of the World's Genetic Resources," *Global Genetic Resources: Access, Ownership and Intellectual Property Rights*, K. E. Hoagland and A. Y. Rossman (eds.) Beltsville symposia in Agricultural Research. Washington, DC: The Association of Systematics Collections.

Day-Rubenstein, K., Stuart, M. & Frisvold, G. (2000). "Agricultural land use in tropical forested countries: patterns, determinants, and implications for biodiversity loss," *World Resource Review*, Vol. *12 (4)*.

Duvick, D. N. (1977). "Genetic Rates of Gain in Hybrid Maize Yields During the Past 40 Years," *Maydica*, Vol. *22 (4)*.

Duvick, D. N. (1984). "Genetic Diversity in Major Farm Crops on the Farm and in Reserve," *Economic Botany*, Vol. *38*.

Duvick, D. N. (1986). "Plant Breeding: Past Achievements and Expectations for the Future," *Economic Botany*, Vol. *40*, No. 3.

Duvick, D. N. (1991). "Industry and Its Role in Plant Diversity," *Forum for Applied Research and Public Policy*, Vol. *6*, No. 3.

Duvick, D. N. (1992). "Plant Breeding: Past Achievements and Expectations for the Future," *Economic Botany*. Vol. *40*.

Duvick, D. N. (1998). "The United States," *Maize Seed Industries in Developing Countries*. M.L. Morris (ed.), Boulder, CO: Lynne Rienner Publishers.

Evenson, R. & D. Gollin (1997). "Genetic Resources, International Organizations and Improvement in Rice Varieties," *Economic Development and Cultural Change,* Vol. *45*.

Evenson, R. E. & Lemarié, S. (1998). "Optimal collection and search for crop genetic resources," *Farmers, Gene Banks and Crop Breeding : Economic Analyses of Diversity in Wheat, Maize and Rice*, M. Smale (ed.), London: Kluwer.

Falck Zepeda, J. & Traxler, G. (1997). "*The Role of Federal, State and Private Institutions in Seed Technology Generation in the United States: The Case of Cotton.*" Paper presented at the Economic Research Service/Farm Foundation Workshop on Agricultural R&D, Rosslyn, VA, Aug. 5.

Feyerherm, A. M. & Paulsen, G. M. (1981). "An Analysis of Temporal and Regional Variation in Wheat Yields," *Agronomy Journal*, Vol. *73*.

Feyerherm, A. M., Paulsen, G. M. & Sebaugh, J. L. (1984). "Contribution of Genetic Improvement to Recent Wheat Yield Increases in the USA," *Agronomy Journal*, Vol. *76*.

Fowler, C. (1991). "Stakes High in Battle for Genetic Diversity," *Forum for Applied Research and Public Policy,* Vol. *6*, No. 3.

Frankel, O. H. (1970). "The genetic dangers of the Green Revolution," *World Agriculture*, Vol. *19*, No. 3.

Frisvold, G. & Condon, P. (1994). "Biodiversity Conservation and Biodiversity Development Agreements," *Contemporary Economic Policy*, Vol. *12*.

Frisvold, G., Sullivan, J. & Raneses, A. (2003). "Genetic improvements in major U.S. crops: the size and distribution of benefits," *Agricultural Economics*, Vol. *28*, No. 2 (March).

Fuglie, K., Ballenger, N., Day, K., Klotz, C., Ollinger, M., Reilly, J., Vasavada, U. & Yee, J. (1996). *Agricultural Research and Development: Public and Private Investments Under Alternative Markets and Institutions,* AER-735, U.S. Department of Agriculture, Economic Research Service.

Global Crop Diversity Trust (2005) www.startwithaseed.org/items/ trust.php Accessed March 11, 2005.

Gollin, D. & Evenson, R. (1990). *Genetic Resources and Rice Varietal Improvement in India*, New Haven, CT: Economic Growth Center, Yale University, unpublished manuscript.

Gollin, D., Smale, M. & Skovmand, B. (2000). "Searching an *Ex Situ* Collection of Genetic Resources," *American Journal of Agricultural Economics*, Vol. *82*.

Goodman, M. & Castillo-Gonzalez, F. (1991). "Plant Genetics: Politics and Realities," *Forum for Applied Research and Public Policy*, Vol. *6*.

Gregorius, H. R. (1978). "The Concept of Genetic Diversity and its Formal Relationship to Heterozygosity and Genetic Distance," *Mathematical Biosciences*, Vol. *41*.

Hanemann, W. M. (1988). "Economics and the Preservation of Biodiversity," *Biodiversity*, E. O. Wilson (ed.), Washington, DC: National Academy Press.

Harlan, J. R. (1971). "Agricultural Origins: Centers and Noncenters," *Science*, Vol. *174*.

Harlan, J. R. (1992). *Crops and Man*. Madison, WI: American Society of Agronomy, Inc., and Crop Science Society of America, Inc.

Hartell, J. G. (1996). *The Contribution of Genetic Resource Diversity: The Case of Wheat Productivity in the Punjab of Pakistan*, M.Sc. Thesis, University of Minnesota, St. Paul, MN.

Heisey, P. W., Srinivasan, C. S. & Thirtle, C. (2001). *Public Sector Plant Breeding in a Privatizing World.* AIB-772. U.S. Department of Agriculture, Economic Research Service.

Heisey, P. W., Lantican, M. A. & Dubin, H. J. (2002). *Impacts of International Wheat Breeding Research in Developing Countries, 1966-97*. Mexico, D.F.: CIMMYT.

Houghton, R. A. (1994). "The Worldwide Extent of Land-use Change," *BioScience*, Vol. *44*, No. 5.

Kaplan, J. K. (1998). "Conserving the World's Plants," *Agricultural Research*, Vol. *46*, no. 9.

Kempthorne, O. (1969). *An Introduction to Genetic Statistics.* Ames, IA: Iowa State University.

Keystone Center (1990). *Final Consensus Report of the Keystone International Dialogue Series on Plant Genetic Resources, Madras Plenary Session.* Washington, DC: Genetic Resources Communications Systems.

Keystone Center (1991). *Final Consensus Report: Global Initiative for the Security and Sustainable Use of Plant Genetic Resources: Keystone International Dialogue Series on Plant Genetic Resources, Oslo Plenary Session.* Washington, DC: Genetic Resources Communications Systems.

Kloppenburg, J. R. (1988). *First the Seed: The Political Economy of Plant Biotechnology*, Cambridge: Cambridge University Press.

Knudson, M. (1998). "Agricultural Diversity: Do We Have the Resources to Meet Future Needs?" *Global Environmental Change: Assessing the Impacts.* G. Frisvold and B. Kuhn (eds.). Aldershot: Edward Elgar Publishing.

Koo, B., Pardey, P. G. & Wright, B. D. with Bramel, P., Debouck, D., van Dusen, M. E., Jackson, M. T., Rao, N. K., Skovmand, B., Taba, S. & Valkoun, J. (2004). *Saving Seeds: The Economics of Conserving Genetic Resources Ex Situ in the Future Harvest Centres of the CGIAR.* Wallingford, UK: CABI Publishing.

Koo, B., Pardey, P. G. & Wright, B. D. (2002). *Endowing Future Harvests: The Long-Term Costs of Conserving Genetic Resources at the CGIAR* Centres, International Plant Genetic Resources Institute, Rome.

Kronstad, W. E. (1996). "Genetic Diversity and the Free Exchange of Germplasm in Breaking Yield Barriers," *Increasing Yield Potential in Wheat: Breaking the Barriers.* M.P. Reynolds, S. Rajaram, and A. McNab (eds.), Mexico, D.F.: CIMMYT.

Magurran, A. (1991). *Ecological Diversity and Its Measurement.* Princeton, NJ: Princeton University Press.

Malecot, G. (1948). *Les Mathematiques de l'Hérédité.* Paris: Masson.

Meng, E.C.H., M. Smale, M. Bellon and D. Grimanelli (1998). "Definition and Measurement of Crop Diversity for Economic Analysis," *Farmers, Gene Banks and Crop Breeding: Economic Analyses of Diversity in Wheat, Maize, and Rice.* M. Smale (ed.) Boston, MA: Kluwer Academic Publishers.

Mooney, P. (1979). *Seeds of the Earth: a Private or Public Resource?* Ottawa:

Canadian Council for International Cooperation.

Mooney, P. (1983). "The Law of the Seed: Another Development and Plant Genetic Resources," *Development Dialogue*, Vol. *12*.

Morris, M. L. (2002). *Impacts of International Maize Breeding Research in Developing Countries*, 1966-98. Mexico, D.F.: CIMMYT.

Morris, M. L. & Heisey, P. W. (2003). "Estimating the benefits of plant breeding research: methodological issues and practical challenges," *Agricultural Economics, 29*.

Morris, M. L., Risopoulos, J. & Beck, D. (1999). *Genetic Change in Farmer-Recycled Maize Seed: A Review of the Evidence*. CIMMYT Economics Program Working Paper 99-07. Mexico, D.F.: CIMMYT.

National Research Council (NRC) (1972). *Genetic Vulnerability of Major Crops*. Washington, DC: National Academy of Sciences.

National Research Council (NRC) (1991). *Managing Global Genetic Resources: The U.S. National Plant Germplasm System*, Committee on Managing Global Genetic Resources: Agricultural Imperatives, Board on Agriculture, National Research Council. Washington, DC: National Academy Press.

National Research Council (NRC) (1993). *Managing Global Genetic Resources: Agricultural Crop Issues and Policies*. Committee on Managing Global Genetic Resources: Agricultural Imperatives, Board on Agriculture, National Research Council. Washington, DC: National Academy Press.

Nei, M. (1972). "Genetic Distance Between Populations," *American Naturalist*, Vol. *106*, pp. 283-292.

Nightingale, K. J. (1996). *Trends in Genetic Diversity of Spring Wheat in the Developing World*. East Lansing, MI: M.Sc. Thesis, Michigan State University.

Office of Technology Assessment, United States Congress (OTA) (1987). *Technologies to Maintain Biological Diversity*, OTA-F-330.

Pardey, P. G., Alston, J. M., Christian, J. E. & Fan, S. (1996). *Hidden Harvest: U.S. Benefits from International Research Aid*. Food Policy Report. Washington, DC: International Food Policy Research Institute.

Pardey, P. G., Koo, B., Wright, B. D., van Dusen, M. E., Skovmand, B. & Taba, S. (2001). "Costing the Conservation of Genetic Resources: CIMMYT's *Ex Situ* Maize and Wheat Collection," *Crop Science* Vol. *41*, No. 4.

Pardey, P. G., Koo, B., van Dusen, M. E., Skovmand, B., Taba, S. & Wright, B. D. (2004). "CIMMYT Genebank," *Saving Seeds: The Economics of*

*Conserving Genetic Resources Ex Situ in the Future Harvest Centres of the CGIAR.* Wallingford, UK: CABI Publishing.

Plucknett, D., Smith, N. J. H., Williams, J. & Anishetty, N. (1987). *Genebanks and the World's Food.* Princeton, NJ: Princeton University Press.

Porceddu, E., Ceoloni, C., Lafiandra, D., Tanzarella, O. A. & Scarascia Mugnozza, G. T. (1988). "Genetic resources and plant breeding: problems and prospects," *The Plant Breeding International, Cambridge Special Lecture.* Cambridge, UK: Institute of Plant Science Research.

Pray, C. E. (1990). "Intellectual Property Rights on Plants: Enforcement, R&D and Potential Importance for LDCs." Mimeo. Rutgers University, New Brunswick, NJ.

Pray, C. E. & Knudson, M. (1994). "Impact of Intellectual Property Rights on Genetic Diversity: A Case Study of U.S. Wheat," *Contemporary Policy Issues* (Jan.).

Pray, C. E., Knudson, M. & Masse, L. (1993). "Impact of Changing Intellectual Property Rights on U.S. Plant Breeding R&D." Paper presented at NC204 Conference on U.S. Agricultural Research. Orlando, FL. March.

Rausser, G. & Small, A. (2000). "Valuing research leads: bioprospecting and the conservation of genetic resources," *Journal of Political Economy,* Vol. *108*, No. 1.

Reid, W. R. (1992). *Genetic Resources and Sustainable Agriculture: Creating Incentives for Local Innovation and Adaptation.* Biopolicy International Series No. 2. Nairobi, Kenya: ACTS Press.

Rejesus, R., van Ginkel, M. & Smale, M. (1996). "Wheat Breeders' Perspectives on Genetic Diversity and Germplasm Use: Findings from an International Survey," *Plant Varieties and Seeds,* Vol. *9*, No. 3.

Reynolds, J., Weir, B. S. & Cockerham, C. C. (1983). "Estimation of the Coancestry Coefficient: Basis for a Short-Term Genetic Distance," *Genetics* Vol. *105*.

Schmidt, J. W. (1984). "Genetic Contributions to Yield Gains in Wheat," *Proceedings of a Symposium on Genetic Contributions to Yield Gains of Five Major Crop Plants*, W.R. Fehr (ed.). Sponsored by C-1, Crop Science Society of America, Atlanta, GA, Dec. 2, 1981. Crop Science Society of America and American Society of Agronomy, Madison, WI.

Sedjo, R. (1992). "Property Rights, Genetic Resources, and Biotechnological Change," *Journal of Law and Economics*, Vol. *35*.

Shands, H. L. (1994). "Some Potential Impacts of the Convention on Biological Diversity on the International System of Exchange of Food

Crop Germplasm," *Conservation of Plant Genetic Resources and the U.N. Convention on Biological Diversity*, D. Witmeyer and M. Strauss (eds.). Washington, DC: American Association for the Advancement of Science.

Shands, H. L. & Stoner, A. K. (1997). "Agricultural Germplasm and Global Contributions," in *Global Genetic Resources: Access, Ownership and Intellectual Property Rights,* K. E. Hoagland and A. Y. Rossman (eds.). Beltsville Symposia on Agricultural Research, Washington, DC: The Association of Systematics Collections.

Siebeck, W. E. (1994). *Personal communication.* Special Advisor to the Vice President for Development Economics, The World Bank, Washington, DC.

Simpson, D. R. & Sedjo, R. A. (1998). "The value of genetic resources for use in agricultural improvement," *Agricultural Values of Plant Genetic Resources,* R.E. Evenson, D. Gollin and V. Santaniello (eds.). Wallingford, UK, CABI Publishing.

Simpson, D. R., Sedjo, R. A. & Reid, J. W. (1996). "Valuing Biodiversity for Use in Pharmaceutical Research," *Journal of Political Economy*, Vol. *104*, No. 1.

Simpson, D. R. & Sedjo, R. A. (1992). "Contracts for Transferring Rights to Indigenous Genetic Resources," *Resources*, Fall.

Smale, M. (1997). "The Green Revolution and Wheat Genetic Diversity: Some Unfounded Assumptions," *World Development*, Vol. *25*, No. 8.

Smale, M. (ed.) (1998). *Farmers, Gene Banks and Crop Breeding: Economic Analyses of Diversity in Wheat, Maize, and Rice.* Boston, MA: Kluwer Academic Publishers.

Smale, M., Reynolds, M. P., Warburton, M., Skovmand, B., Trethowan, R., Singh, R. P., Ortiz-Monasterio, I., Crossa, J., Khairallah, M. & Almanza-Pinzon, M. I. (2001). *Dimensions of Diversity in CIMMYT Bread Wheat from 1965 to 2000*. Mexico, D.F.: CIMMYT.

Smale, M. & McBride, T. (1996). "Understanding Global Trends in the Use of Wheat Diversity and International Flows of Wheat Genetic Diversity," Part 1 of *CIMMYT 1995/6 World Wheat Facts and Trends 1995/6: Understanding Global Trends in the Use of Wheat Diversity and International Flows of Wheat Genetic Diversity*. Mexico, D.F.: CIMMYT.

Smale, M., Hartell, J., Heisey, P. & Senauer, B. (1998). "The Contributions of Genetic Resources and Diversity to Wheat Production in the Punjab of Pakistan," *American Journal of Agricultural Economics*, Vol. *80*.

Smale, M. & Day-Rubenstein, K. (2002). "The Demand for Crop Genetic Resources: International Use of the U. S. National Plant Germplasm

System," *World Development*, Vol. *30*, No. 9.

Solow, A., Polasky, S. & Broadus, J. (1993). "On the Measurement of Biological Diversity," *Journal of Environmental Economics and Management*, Vol. *24*.

Souza, E., Fox, P., Byerlee, D. & Skovmand, B. (1994). "Spring Wheat Diversity in Irrigated Areas of Two Developing Countries," *Crop Science*, Vol. *34*.

Sperling, C. (1994). Personal communication. United States Department of Agriculture, National Germplasm Resources Laboratory, Beltsville, MD.

Swaminathan, M. (1996). "Compensating Farmers and Communities through a Global Fund for Biodiversity Conservation for Sustainable Food Security," *Diversity*, Vol. *12*.

Swanson, T. (1996). "Global Values of Biological Diversity: The Public Interest in the Conservation of Plant Genetic Resources of Agriculture," *Plant Genetic Resources Newsletter No. 105*. Rome: International Plant Genetic Resources Institute.

Tessier, N. & Bernatchez, L. (1999). "Stability of Population Structure and Genetic Diversity Across Generations Assessed by Microsatellites Among Sympatric Populations of Landlocked Atlantic Salmon (*Salmo salar* L.)," *Molecular Ecology*, Vol. *8*.

Thirtle, C. G. (1985). "Technological Change and the Productivity Slowdown in Field Crops: United States, 1939-78," *Southern Journal of Agricultural Economics*, Vol. *17* (Dec.).

Tilman, D. (1998). "The greening of the Green Revolution," *Nature*, Vol. *396*.

U.N. FAO (1996a). *Report of the International Technical Conference on Plant Genetic Resources*. Rome.

U.N. FAO (1996b). *Report on the State of the World's Plant Genetic Resources*. Rome.

U.N. FAO (1996c). *Global Plan of Action for the Conservation and Sustainable Utilization of Plant Genetic Resources for Food and Agriculture, and the Leipzig Declaration*. Adopted by the International Technical Conference on Plant Genetic Resources, Leipzig, Germany, June 17-23, Rome.

U.N. FAO (1997). *Revision of the Cost Estimates of the Global Plan of Action*. Rome.

U.N. FAO (1998). *The State of the World's Plant Genetic Resources for Food and Agriculture*. Rome.

U.S. Department of Agriculture (1990). "New Germplasm Helps Assure Food for the Future," USDA News Feature, News Division, Office of Public

Affairs, Dec. 6.

U.S. Department of Agriculture, Economic Research Service (2002). ERS-USDA Land Use Database. www.ers.usda.gov/Briefing/LandUse/agchangechapter.htm#regional

United States Department of State (1994). Correspondence between Warren Christopher, Secretary of State, and Senator Robert Dole, Aug. 16.

United States General Accounting Office (GAO) (1997). *U. S. Department of Agriculture: Information on the Condition of the National Plant Germplasm System*. Report to Congressional Committees.

Webster, P. (2003). "Prestigious Plant Institute in Jeopardy," *Science,* Vol. *229* (Jan. 31).

Weitzman, M. L. (1992). "On Diversity," *Quarterly Journal of Economics* Vol. *107.*

Weitzman, M. L. (1993). "What to Preserve? An Application of Diversity Theory to Crane Conservation," *Quarterly Journal of Economics*, Vol. *108.*

Widawsky, D. A. (1996). *Rice Yields, Production Variability, and the War Against Pests: An Empirical Investigation of Pesticides, Host-Plant Resistance, and Varietal Diversity in Eastern China*. Palo Alto, CA: Ph.D. dissertation, Stanford University.

Wilkes, G. (1991). "*In Situ* Conservation of Agricultural Systems," *Biodiversity: Culture, Conservation, and Ecodevelopment,* J. B. Alcorn and M. Oldfield, (eds.) . Boulder, CO: Westview Press.

World Resources Institute (WRI), United Nations Environment Programme, United Nations Development Programme, and The World Bank (1991). *World Resources, 1990-1991.* New York: Oxford University Press.

Wright, S. (1922). "Coefficients of Inbreeding and Relationship," *American Naturalist,* Vol. *56.*

Zohrabian, A. (1995). *Genetic Resources Conservation and Maintenance in the Former Soviet Union.* Department of Agricultural and Applied Economics Staff Paper P951. Minneapolis: University of Minnesota.

## End Notes

[1] Genetic resources may also have economic value even if they are not currently being used. By preserving resources, we retain the option to use them in the future, when they may become important for agricultural, pharmaceutical, ecological, or industrial applications—even if we do not currently know precisely what those resources or applications are (Kaplan, 1998). Even if they are never used, diverse genetic resources may be valued by some people

simply for their existence, or as a bequest left intact to future generations (Barbier et al., 1995).

[2] Growth accounting is often indicative rather than exact. Various factors (such as improved germplasm and improved crop management practices) frequently interact with one another, making it difficult to isolate the contributions of a single source. Interaction also means that the productivity gain from simultaneous adoption often exceeds the sum of the productivity gains when new varieties or crop management practices are adopted separately (Morris and Heisey, 2003).

[3] See Duvick (1977, 1984, 1992) on maize (corn) in the U.S., and Feyerherm and Paulsen (1981); Feyerherm, Paulsen, and Sebaugh (1984); Schmidt (1984); Cox et al. (1988); and others listed by Heisey, Lantican, and Dubin (2002) on wheat in both industrialized and developing countries.

[4] Most studies have focused only on genetic improvements for a single crop.

[5] The average annual growth in U.S. crop yields during 1975–92 was 1.33 percent for corn, 1.54 percent for sorghum, 1.13 percent for wheat, 1.23 percent for soybeans, and 2.23 percent for cotton. The half of the yield growth not attributed to improved seed varieties came from other inputs and management factors, including more fertilizers and pesticides, better agronomic practices, and investments in irrigation and drainage. These other sources of productivity growth also may have been affected by agricultural research.

[6] In other words, annual yield gains attained in a given year are maintained in the years following.

[7] Of course in the long term, research gains may be counteracted by losses of resistance to pests and diseases, but in a successful research program the net gains are positive. The point here is that it is more realistic to look at research gains as a permanent stream over time rather than as an economic benefit occurring only once. If investments are ongoing, new additions to the permanent stream are received every year. Furthermore, avoidance of losses is in fact an economic benefit as well.

[8] Koo et al. present additional cost figures for CGIAR gene banks. For many cost compoinents of gene bank operation, cost estimates fall between the estimates for wheat and maize.

[9] For example, certain isolated areas in Mediterranean Europe grow wheat landraces. Faro, or Triticum dicoccum, is grown in Italy.

[10] This refers to countries for which a large proportion of the corn produced enters the formal market.

[11] In another example, modern corn hybrids adapted to the Midwestern United States were derived from dent varieties from the Southeastern United States and flint varieties from the Northeast, which were themselves adapted by settler farmers from many locally distinct varieties selected and reselected by Native American farmers over many previous generations (Duvick, 1998)

[12] After the breakup of the Soviet Union, the Vavilov Institute, one of the largest collections in the world, has faced critical financial and structural problems. Funding for gene banks in Russia and in these republics has been greatly reduced and many accessions are at risk (Zohrabian, 1995; Webster, 2003).

[13] A GAO study of the U.S. Plant Germplasm System echoed the concerns of the FAO report (GAO, 1997).

[14] Goodman and Castillo-Gonzalez (1991) also note that "improved breeding lines have been less freely exchanged, even among public agencies."

[15] Unshielded genetic materials also contain improved varieties; in fact, improved breeding materials are the type of germplasm most frequently distributed by the U.S. National Plant Germplasm System. While public research institutions are the primary source of germplasm placed in the NPGS, private breeding concerns donate materials as well, particularly obsolete breeding materials.

[16] Utility patents are the broadest class of patents and, unlike plant patents, they can be used for sexually reproducing plants.

[17] The United States signed the Convention in June 1993, but the U.S. Senate has not yet ratified it

[18] Other aspects of the MTAs, such as recordkeeping and means of assigning parentage of a variety, have yet to be worked out in detail.

[19] See Smale (1998), and particularly the chapter by Meng et al. (1998), for one of the first attempts to summarize the application of various diversity- related measures to crops and to give these measures an economic interpretation.

[20] Another characteristic, infrequently noted, of both morphological and genetic measures is that they obviously require informed choice of the characteristics or genes that will be analyzed. No index will be constructed, for example, based on all genes in a crop that are polymorphic, i.e., genes that have more than one variant. In the first place, such a list is unknown, and in the second, costs would become completely prohibitive.

[21] This approach uses the coefficient of diversity (COD), which equals 1 - the coefficient of parentage (COP). The COP is a pairwise comparison based on pedigree analysis (Wright, 1922; Malecot, 1948; Kempthorne, 1969; Cox et al., 1985). COD/COP analysis is less costly than analysis of proteins or molecular methods, but it also has some disadvantages: 1) it ignores the possibility that alleles could be identical even without common heritage; 2) it relies on the assumption that the ultimate ancestors that are recorded in a pedigree are unrelated, which may not be true; and 3) it assumes that "each parent contributes equally to offspring, despite the effects of recurrent selection and random genetic drift" (See Nightingale, 1996; Cox et al., 1985; Meng et al., 1998)

In: Plant Genetic Resources and Food Security ISBN: 978-1-61761-732-4
Editors: Marlena A. Diaz 

*Chapter 3*

# USING GENETIC TOOLS TO COMBAT HUNGER

***Kay Simmons and Steven M. Kappes***

Walk into any grocery store and you will see it for yourself: We are producing an unprecedented bounty of food. Having such an abundant food supply begs the question: Why work so hard at improving our crops and livestock when we are already so successful? The answer is simple. We live in a changing world.

The world's population, now at 6.8 billion people, has more than doubled since the 1950s and is expected to reach 9 billion by 2050. The United Nations Food and Agriculture Organization predicts that food production will need to double by 2050 to meet the increased demand. Water supplies will also be a concern as the need to irrigate crops competes with demands from thirsty cities and suburbs in places as diverse as Beijing, New Delhi, and Phoenix.

Climate change is altering landscapes in ways we are only beginning to understand, affecting air temperatures, rainfall patterns, soil dynamics, and the seasonal cycles so vital for a bountiful harvest. Some experts predict that warmer temperatures will reduce yields and cause global food shortages.

New threats from pests and pathogens are also emerging. Ug99, a fungal pathogen, has become an international threat to wheat supplies since its discovery was reported in Uganda a decade ago. Sheath blight, considered the world's worst rice pathogen, has emerged as more of a danger since the 1970s, when scientists developed higher yielding rice varieties.

Chemical and agronomic solutions to pest, weed, and pathogen problems continue to evolve. Some research taps into the genetics and physiology of mosquitoes, ticks, and other pests to find environmentally sound treatments that will target specific arthropods by exploiting how they breathe, feed, shed cells, and reproduce.

When we talk about food supplies, we need to consider livestock health as well as human and crop health. For example, to subsistence farmers in sub-Saharan Africa and many other developing areas, bovine diseases can mean the difference between success and starvation by threatening just a few head of cattle.

To address these challenges, scientists are deciphering the DNA of our most important crops and livestock and tapping into genes that offer enhanced nutritional value, increased resistance to pests and diseases, and the ability to survive in changing climates. ARS researchers have been leading the way, unlocking genetic clues that have been instrumental in the development of beef and dairy cattle that are more productive and varieties of wheat, rice, corn, beans, and potatoes that are hardier and more nutritious.

For example, scientists in Stuttgart, Arkansas, are using DNA markers to identify rice varieties with genetic resistance to sheath blight. Other teams—in Beaumont, Texas, and New Orleans, Louisiana—are using rice genes to unlock nutrients in a new variety of high-fiber rice that may create a buzz with its distinct purple color.

Much of the research has an international reach. ARS scientists in Stoneville, Mississippi, are working with colleagues in Paraguay to identify genes that resist Asian soybean rust, a worldwide threat to soybeans. In Beltsville, Maryland, scientists are broadening the genetic base of beans to identify genes that resist the rusts that damage harvests in Africa and the Americas.

Another Beltsville team is working with farmers in Africa to breed hardier and more productive cattle by using technology developed by ARS scientists with help from international colleagues. The Illumina Bovine SNP50 BeadChip, a glass slide containing thousands of DNA markers, can identify useful genetic traits and has already proved to be a key tool in the United States for genotyping bulls that will sire offspring with desirable milk production traits. To address the threat posed by Ug99, ARS researchers at several locations are collaborating with scientists in Kenya and at the International Maize and Wheat Improvement Center in Mexico to explore the genetics of both wheat and the pathogen.

In Aberdeen, Idaho, ARS scientists have produced potato varieties with increased protein and vitamin C content and are collaborating with Mexican scientists on field trials designed to find potatoes that resist late blight fungus, the pathogen that caused the Irish potato famine.

In some areas of sub-Saharan Africa, people get up to 60 percent of their calories from corn, which is very low in vitamin A. This diet can lead to vitamin A deficiencies that cause infant mortality, eye diseases, and blindness among children. ARS scientists in Ithaca, New York, discovered two varieties of corn that could increase vitamin A levels 15-fold. And researchers at the Children's Nutrition Research Center in Houston, Texas, have shown that Golden Rice-2, a variety 20 years in the making, will be effective at fighting vitamin A deficiencies.

We must grow our food smarter, with less water and on landscapes altered by climate change and threatened by evolving diseases and pests. ARS scientists are addressing that challenge, using genetics to develop crops and livestock that are more resilient and more nutritious. The work is a necessity not only for our health, but also for our survival in a changing world.

**Kay Simmons**
ARS Acting Deputy Administrator
Crop Production and Protection
Beltsville, Maryland

**Steven M. Kappes**
ARS Deputy Administrator
Animal Production and Protection
Beltsville, Maryland

In: Plant Genetic Resources and Food Security ISBN: 978-1-61761-732-4
Editors: Marlena A. Diaz 

*Chapter 4*

# RICE: IMPROVING RICE, A STAPLE CROP WORLDWIDE

## *Agricultural Research Forum*

Rice *(Oryza sativa)* is the main dietary staple for more than half the world's population. In 2008, worldwide rice consumption exceeded 430 million metric tons. But the world's continued rice supply is jeopardized by a myriad of factors, including diseases and inability to keep up with demand.

Since the Agricultural Research Service is a world leader in rice research, it's no surprise that scientists at the Dale Bumpers National Rice Research Center in Stuttgart, Arkansas, and the Rice Research Unit in Beaumont, Texas, are involved in domestic and international efforts to improve rice varieties worldwide.

## GIVING SHEATH BLIGHT THE GENETIC SHOVE

As part of a multi-state and multi- institutional project called "RiceCAP," supported by the USDA National Institute of Food and Agriculture, geneticists Anna McClung and Georgia Eizenga and plant molecular pathologist Yulin Jia, at Stuttgart, and geneticist Shannon Pinson, at Beaumont, are working to pinpoint the exact genes that confer resistance to sheath blight. The disease is caused by *Rhizoctonia solani,* a fungus that can stay in the soil for 20 years, killing the plant's cells each planting cycle and affecting yield and grain quality.

The researchers had a breakthrough in their sheath blight mapping efforts when they identified and confirmed that a particular region on a chromosome—known as "quantitative trait loci *qShB9-2* "—has a major effect on controlling the disease.

"This is our most important discovery thus far on this project," says Jia. "It's the first time we have found—and are confident in—a chromosomal region with genetic resistance to this pathogen. It will now be easier to develop resistant rice varieties."

In a related project, Eizenga and colleagues screened wild rice species for signs of sheath blight resistance. Of the 73 wild species obtained from the International Rice Research Institute (IRRI) in the Philippines, 7 accessions showed the most resistance. The scientists have crossed some of these accessions with U.S. varieties lacking resistance, hoping to transfer resistance genes from these wild species to cultivated rice to create new germplasm. The researchers are currently working on mapping populations to identify the exact locations of these resistance genes.

The Stuttgart group has also developed a standardized greenhouse-screening technique for accurate phenotyping of sheath blight. In this technique, called the "microchamber method," 2-liter or 3-liter soft drink bottles are used to create a humidity chamber to promote disease development, allowing the scientists to measure the rice seedlings' disease reaction in just 7 days. This has accelerated the overall process of identifying novel, resistant resources from cultivated and wild relatives of rice. The technology has since been transferred to several rice-research programs in Latin America to help countries evaluate their rice varieties for sheath blight and other rice diseases.

Meanwhile, in Beaumont, Pinson and colleagues have been studying gene- mapping populations developed from recombinant inbred lines (RILs) of Lemont a domestic cultivar, and TeQing, a cultivar from China. Because each of these lines contains different combinations of DNA, they can be used efficiently to find chromosomal regions containing genes for resistance to sheath blight. Pinson has been able to find 18 chromosomal regions with genes that help plants resist damage from the disease. Two of those chromosomal regions have shown a large, measureable effect on sheath blight resistance and were associated with flowering time and plant height.

Technician Piper Roberts (left) and geneticist Shannon Pinson look for disease symptoms after inoculation at the sheath blight disease nursery at Beaumont, Texas.

A rice leaf exhibiting typical watermark lesions associated with sheath blight disease.

In a Stuttgart, Arkansas, rice field, ARS plant molecular pathologist Yulin Jia (left) evaluates sheath blight disease in a mapping population with his assistant Guangjie Liu, of the University of Arkansas.

The Lemont/TeQing RIL gene-mapping population is uniquely important in that it has two high-yielding cultivars as parents, is the first to be well adapted to the southern United States and other semitropical growing conditions, and contains more progeny lines than other rice gene-mapping populations, which makes estimation of genetic marker locations more precise and evaluations of marker-linkages easier for breeders.

Pinson has also teamed up with a scientist in the Philippines to develop a second gene-mapping population known as the "TeQing-into-Lemont introgression lines" (TILs), which consist of 123 genetic lines that each contain just 1 to 5 small pieces of TeQing DNA placed within a predominantly Lemont genetic background. This consistent genetic background surrounding the foreign TeQing genes allows researchers to more accurately measure the effect of each small piece of TeQing DNA. By looking at the trait expression of the TILs compared with each other and with pure Lemont plants, scientists can find genes having a small but significant impact and can determine their genomic location more precisely, which could help researchers improve worldwide rice production.

## TAKING THE BANG OUT OF RICE BLAST

Rice blast, caused by the fungus *Magnaporthe oryzae,* is a disease threatening rice worldwide. The fungus is airborne but is also transmitted through seed, infecting rice plants during all developmental stages. Strains of rice blast are always changing, making it a challenge to continually produce varieties resistant to it.

But McClung, Jia, fellow geneticist Bob Fjellstrom, and colleagues have made a breakthrough in understanding the disease. The scientists have developed molecular markers to screen for resistance genes. They have also found the *Pi-ta* gene gives rice resistance to several races of rice blast.

"For a long time, scientists believed that one gene only produces one protein to prevent infection by one race of the blast fungus," says Jia. "Our studies show that one gene may produce multiple proteins. *Pi-ta* can make 12 proteins, each capable of conferring resistance to up to 10 races. We have identified those races, giving breeders valuable information to use when selecting parents of new cultivars."

Recently, rice blast race IE-1k—a race to which *Pi-ta* doesn't confer resistance— appeared in rice paddies in the southern United States. Jia and colleagues found genes *Pi42(t)* and *Pi43(t)*, from the Chinese cultivar Zhe733, confer resistance not only to IE-1k, but also to all common races of rice blast found in the United States. According to Jia, breeders can overlap *Pi-ta*, *Pi42(t)*, and *Pi43(t)* to increase cultivars' longevity of resistance to rice blast.

## STANDARDIZED AMYLOSE TESTING IMPROVES RICE

In addition to increasing yield and disease resistance, rice breeders try to ensure that the amylose content of a cultivar fits into a range that will provide a certain level of cooking quality. Starch amylose content is the key factor affecting the texture and processing properties of rice.

Using the microchamber method, Yulin Jia evaluates sheath blight disease of rice seedlings a week after inoculation with the fungus *Rhizoctonia solani*.

Geneticist Melissa Jia loads plates containing DNA from a sheath blight mapping population into an automated genotyping system while Guangjie Liu, of the University of Arkansas, analyzes the output to identify genes associated with sheath blight resistance.

And recent studies have correlated amylose content to "resistant starch"—which, after consumption, resists digestion in the small intestine. Resistant starch has been proposed to have health benefits similar to those of dietary fiber.

Amylose levels help rice breeders decide how best to market a rice variety. The amylose content typical of U.S. long-grain rice ranges from 19 to 23 percent, giving it a dry, fluffy texture after cooking. The typical U.S. medium-grain rice, which has an amylose content of less than 19 percent, is soft and sticky after cooking. Rice with an amylose content higher than 23 percent is used for canning.

"When rice breeders ask about the grain quality traits of a possible new rice germ- plasm source, amylose content is one of the traits that is looked at carefully," says Beaumont chemist Ming-Hsuan Chen.

But there is a problem: The reported amylose content might not actually be the amount that the breeder is looking for. This is because the test to determine amylose content is not standardized worldwide, so test results vary from country to country—even for the exact same variety. For example, in 2009 the rice cultivars Kyeema and Doongara were reported to have amylose contents of 14 and 22 percent, respectively, in one country, 19 and 24 percent in another, and 20 and 28 percent in yet another.

A diseased rice leaf 1 week after inoculation with the rice blast fungus *Magnaporthe oryzae* under greenhouse conditions at Dale Bumpers National Rice Research Center, Stuttgart, Arkansas.

"Such differences make it particularly difficult to exchange germplasm between breeding programs, present data to international audiences, or make associations between amylose values and traits or genotypes," says Chen.

Each year, thousands of rice samples are analyzed at the Beaumont laboratory for amylose content and other critical end- use quality traits. The scientists there have standardized their method of determining amylose content so that the values are repeatable year after year.

## GOLDEN RICE-2 SHINES IN NUTRITION STUDY

All across America, rice has a loyal following among those who enjoy crispy rice cereal at breakfast, steamed white rice with a favorite entree at lunch, or a classic rice pudding as an evening dessert.

But America's consumption of rice—about 21 pounds per person each year—is substantially less than that of people who live in the world's "rice-eating regions," mainly Asia, most of Latin America, and much of Africa.

Because vitamin A deficiency—and its harmful impacts on health—is common in some of these overseas areas, scientists in Europe and the United States have worked for more than a decade to genetically engineer white rice so that it will provide beta-carotene. Our bodies convert beta-carotene into retinol, a form of vitamin A.

White rice typically does not have any detectable beta-carotene. But the genetically engineered Golden Rice-2 from Syngenta Corporation does. Until now, however, scientists haven't known how efficiently our bodies can convert the beta-carotene in Golden Rice-2 into retinol.

Research published in a 2009 issue of the *American Journal of Clinical Nutrition* provides a scientifically sound answer. Agricultural Research Service plant physiologist Michael A. Grusak, carotenoids researcher Guangwen Tang, and colleagues reported, for the first time, their findings that one 8-ounce cup of cooked Golden Rice-2 provides about 450 micrograms of retinol. That's 50 to 60 percent of the adult Recommended Dietary Allowance of vitamin A.

Tang, who led the study, is at the ARS Jean Mayer USDA Human Nutrition Research Center on Aging at Tufts University, Boston, Massachusetts; Grusak is with the ARS Children's Nutrition Research Center at Baylor College of Medicine in Houston, Texas.

ARS, the National Institutes of Health, and the U.S. Agency for International Development funded the research.

The scientists based their determinations on tests with five healthy adult volunteers who ate one serving of the rice at the start of the 36-day study. Volunteers' blood was sampled at more than 30 intervals during the research. By analyzing those samples, the researchers were able to determine the amount of beta-carotene (and retinol) that the volunteers absorbed (and then converted to retinol) from the Golden Rice-2.

The efficient conversion of Golden Rice-2 beta-carotene into vitamin A strongly suggests that, with further testing, this special rice might help reduce the incidence of preventable night blindness and other effects of vitamin A deficiency in rice-eating regions. Right now, more than 200 million people around the globe don't get enough vitamin A.

Grusak conducted experiments that made it possible for Tang's group to detect beta-carotene (and resultant retinol) derived from Golden Rice-2, differentiating it from beta-carotene or retinol from other sources.

In his experiments, Grusak determined how to get Golden Rice-2 plants, grown in his rooftop greenhouse at Houston, to take up a harmless tracer and incorporate it into the beta-carotene in the developing grains. The tracer, a rare yet safe and natural form of hydrogen, can be detected by a gas chromatograph-mass spectrometer, the kind of instrument that Tang's team in Boston used to analyze volunteers' blood samples.

The tracer, deuterium oxide, is not new to vitamin A research. But Grusak's studies are the first to show how the tracer can be successfully incorporated into the grains of a living plant for vitamin A investigations.

"It was tricky to determine how much tracer to use and when to add it to the nutrient solution we grew the plants in," says Grusak. His method might be used in other pioneering research geared to boosting the nutritional value of other grains worldwide.—By Marcia Wood, ARS.

Golden Rice-2 plants growing in a greenhouse.

*This research is part of Human Nutrition, an ARS national program (#107) described at* www.nps.ars.usda.gov.

*Michael A. Grusak is with the USDAARS Children's Nutrition Research Center at Baylor College of Medicine, 1100 Bates St., Houston, TX 77030, (713) 798-7044,* mike.grusak@ars.usda.gov.

*Guangwen Tang is with the ARS Jean Mayer USDA Human Nutrition Research Center on Aging at Tufts University, 711 Washington St., Boston, MA 02111, (617) 556-3236, guangwen.tang@tufts.edu.*

Chen has joined the International Network for Quality Rice and collaborates on a project led by IRRI's Melissa Fitzgerald and commissioned by the International Union of Pure and Applied Chemistry. The goal of this project is, through worldwide collaboration among rice quality labs, to establish a standardized test for amylose content determination internationally.

## Continued Collaboration Vital to Success

"The exchange of plant germplasm and genetic stocks helps to identify genes and genetic markers that can be used by rice breeders globally to develop new cultivars that will sustain agriculture and help feed the world," says McClung, research leader of both the Stuttgart and Beaumont laboratories.

Scientists in Stuttgart received several wild rice species collected from rice-growing regions around the world and stored at IRRI to help identify novel disease-resistance genes. In return, Stuttgart scientists sent IRRI about 400 purified rice cultivars representative of rice grown around the world, which can be used to identify markers associated with traits important to rice production practices in Asia.

Additionally, ARS, the University of Arkansas, and IRRI have an ongoing informal collaboration to identify genes and plant traits that will contribute to the development of high-yielding rice cultivars with disease resistance. In another project, ARS is working with IRRI and Cornell University to develop 600,000 genetic markers that can be used to identify genes that control yield, grain quality, and resistance to physiological stress, insects, and disease in rice cultivars from around the world.

Such work ensures that people worldwide will be able to enjoy rice for years to come.—By **Stephanie Yao** and **Alfredo Flores,** ARS.

*This research is part of Plant Genetic Resources, Genomics, and Genetic Improvement (#301) and Crop Protection and Quarantine (#304), two ARS national programs described at* www.nps.ars.usda. gov.

*To reach scientists mentioned in this article, contact Stephanie Yao, USDA -ARS Information Staff, 5601 Sunnyside Ave., Beltsville, MD 20705-5129; (301) 504-1619,* stephanie.yao@ars.usda.gov.

In: Plant Genetic Resources and Food Security ISBN: 978-1-61761-732-4
Editors: Marlena A. Diaz 

*Chapter 5*

# BEANS: HELP FOR THE COMMON BEAN-GENETIC SOLUTIONS FOR LEGUME PROBLEMS

## *Agricultural Research Forum*

The common bean—which includes pinto, great northern, navy, black, kidney, and snap beans— is considered by many nutritionists to be a nearly perfect food because of its high protein content and low cost. But it is also susceptible to many diseases that reduce seed and pod quality and yields. Agricultural Research Service scientists from labs across the United States are playing major roles in finding solutions to what ails these legumes.

## BELTSVILLE BEANS KEY TO COMBATING DEVASTATING RUST PATHOGEN

ARS plant pathologist Talo Pastor-Corrales, throughout his career, has traveled to 21 countries in the Americas and 11 in Africa studying bean diseases and searching for bean varieties that contain special traits—particularly disease resistance—that could be used to improve common beans. In the Soybean Genomics and Improvement Research Unit in Beltsville, Maryland, Pastor-Corrales specializes in genetic resistance of the common bean *(Phaseolus vulgaris)* to various diseases.

He's also the lead scientist in a project that aims to discover and breed genes into *P. vulgaris* for resistance to common bean rust and the newly arrived Asian soybean rust pathogen, which also infects the common bean.

The fungus that causes bean rust is very aggressive and exists as many different strains called "races." Pastor-Corrales says, "When new races appear, they can infect bean varieties that were previously resistant to rust." Further complicating matters is the fact that races present in a field can vary from one year to another.

Of major concern is the loss of effectiveness of the *Ur-3* rust-resistance gene in beans, which has been very effective in controlling bean rust in the United States, especially in North Dakota and Michigan, the two largest producers of dry beans in the United States. In recent years, however, rust has developed on these oncerust-resistant bean varieties, and there is concern that the new races will spread to other Northern Plains states, such as Colorado and Nebraska.

In 2008 and 2009, Pastor-Corrales and his project team were credited with developing new dry bean cultivars resistant to the rust pathogen. Pastor-Corrales collaborated with scientists from the University of Nebraska and Colorado State University. The new cultivars contain two or more rust-resistance genes and most also have the *Ur-11* gene, considered the most effective rust-resistance gene in the world.

## Beans That Can Take the Heat

At test plots in southern Puerto Rico, ARS plant geneticist Tim Porch's beans are feeling the heat. As part of collaborative breeding efforts with Cornell University, the University of Nebraska, and the University of Puerto Rico, Porch and colleagues have been testing new bean germplasm for heat and drought tolerance and disease resistance. So far, their efforts have proved fruitful.

Porch is in the process of releasing two new kidney bean varieties with heat tolerance. These germplasm releases, named "TARS HT-1" and "TARS HT-2," were initiated by ARS plant geneticist Rusty Smith, now with the ARS laboratory in Stoneville, Mississippi. TARS HT-1 does well under the stress of high day and high night temperatures, whereas TARS HT-2 does well under the stress of high day and moderate night temperatures.

Also in the works is new black bean germplasm with heat and drought tolerance and resistance to common bacterial blight, a seedborne disease—spread by splashed water—that mainly attacks the plant's leaves and pods. Porch crossed tropical black and red beans to produce these germplasm lines, which are adapted to temperate areas and will help to increase the diversity of U.S. bean germplasm. Field tests in Nebraska show that the lines yield well in addition to having tolerance to heat, drought, and disease.

"The beans we are testing have broad adaptation," says Porch, who is with ARS's Tropical Agriculture Research Station (TARS) in Mayagüez, Puerto Rico. "Our lines do well in the short days common to Puerto Rico and the long days found in Nebraska." Porch is testing other bean types— red, pinto, great northern, and navy—that are drought tolerant, and some also have heat tolerance and disease resistance. "My goal is to pyramid multiple resistances to generate lines with broad adaptation and genetic diversity."

Porch is also involved in bean-improvement efforts in Angola, a country that is beginning to recover from many years of civil war. The project, funded by USAID and led by the University of Puerto Rico, supports Angola's common bean breeding program. Porch and university colleagues conduct breeding and pathology training sessions, host Angolan scientists to train them in the laboratory, and help the scientists breed for traits of importance, such as resistance to angular leaf spot, common bacterial blight, and bean common mosaic virus.

Plant pathologist Talo Pastor-Corrales examines a bean cultivar that is a new source of genes for resistance to a hyper- virulent pathogen that causes rust disease of dry and snap beans.

## WASHINGTON'S WONDERS

At ARS's Vegetable and Forage Crops Research Laboratory in Prosser, Washington, plant pathologist Richard Larsen and geneticist Phil Miklas recently identified new sources of resistance for protecting snap beans from the viral disease chocolate pod, which was first detected in Wisconsin, Michigan, and other Great Lakes states in 2001 and inflicts unsightly defects on pods, ruining their marketability.

Insecticides are sometimes used to kill virus-transmitting aphids. But incorporating resistance into snap beans is considered a more sustainable approach. Toward that end, the researchers devised DNA marker technology to help speed identification and use of plants harboring chocolate pod resistance without having to grow them to maturity.

Reducing insecticide use—and safeguarding the environment—was also the goal of a project that entomologist Stephen Clement recently completed at ARS's Plant Germplasm Introduction and Testing Research Unit in Pullman, Washington. There, as part of a 3-year project supported by the U.S. Agency for International Development, Clement led development of chickpea germplasm lines offering beet armyworm resistance. The moth's caterpillar stage attacks many crops, but is especially problematic in chickpeas in India, which produced 6.6 billion tons of the high-fiber, vitamin-rich crop in 2005.

Clement collaborated on the project with scientists at Washington State University-Pullman and the International Crops Research Institute for the Semi-Arid Tropics (ICRISAT) in Patancheru, India. In U.S. trials, 28 to 62 percent of beet armyworms that fed on the leaves of resistant chickpeas died within a few days. Those that survived were smaller and shorter than usual. Now, ICRISAT entomologist Hari Sharma is conducting field trials to evaluate the resistant chickpeas' potential to forestall insecticide use.

Earlier this year, George Vandemark, a geneticist at ARS's Grain Legume Genetics and Physiology Research Unit, also in Pullman, released a new Eston class lentil named "Essex." This new variety was developed through a collaborative effort involving Vandemark, Fred Muehlbauer (now retired from ARS), and North Dakota State University pulse crop breeder Kevin McPhee.

They chose Essex for release because of its outstanding performance in yield trials conducted in Washington State, Idaho, North Dakota, and Montana—states that produced a combined $87 million's worth of lentils, most of it for export.

On average, Essex yielded 1,220 pounds of seed per acre—21 percent more than Eston and 22 percent more than Athena, commercial varieties used for comparison. Essex is intended for production in the Northern Plains, with Mexico and other Latin American nations as prime export destinations.—By Alfredo Flores, Stephanie Yao, and Jan Suszkiw, ARS.

Geneticist Tim Porch examines the effects of high-temperature stress on pod development in the common bean.

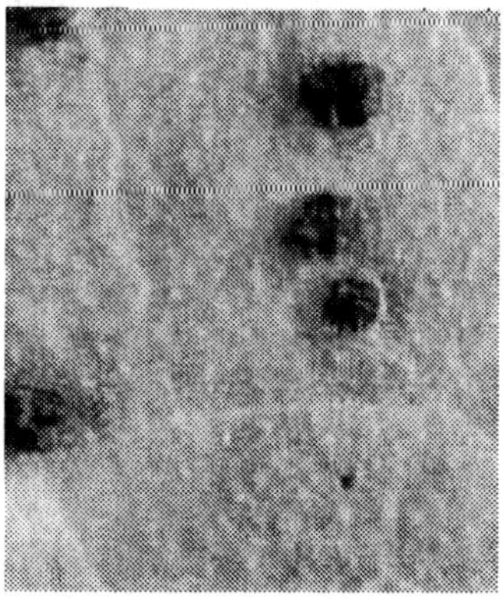

Close-up of pustules (fungal fruiting structures containing thousands of spores) of bean rust fungus on a susceptible bean leaf.

Entomologist Hari Sharma, ICRISAT, and a colleague in a chickpea research plot at the Regional Research Station, Kukumseri, in the Himalayan foothills of northern India.

*This research is part of Plant Genetic Resources, Genomics, and Genetic Improvement (#301) and Plant Diseases (#303), two ARS national programs described at www.nps.ars.usda.gov.*

*To reach the scientists mentioned in this article, contact Alfredo Flores, USDA -ARS Information Staff, 5601 Sunnyside Ave., Beltsville, MD 20705-5129; (301) 504-1627, alfredo.flores@ ars.usda.gov.*

Essex lentils. The average diameter is 7 millimeters.

Geneticist Phil Miklas compares susceptible (left) and resistant (right) bean lines exposed to virus infection.

## BETTER BEANS MEAN BETTER HEALTH *FOR PEOPLE EVERYWHERE*

Whether it's a side of beans for a hearty breakfast, an extra- spicy chili at lunch, or an elegant, chilled black bean soup at dinner, beans can add pleasing color and texture to any meal. And, importantly, beans provide iron, an essential nutrient needed in comparatively small, or "micro," amounts.

In Ithaca, New York, ARS physiologist Raymond P. Glahn, ARS research associate Elad Tako, Cornell University analytical chemist Michael A. Rutzke, and others conduct research that may help plant breeders develop new and improved beans that are even better sources of iron. Their research would especially benefit the more-than-2-billion people around the globe who are deficient in iron. Iron deficiency is, in fact, the world's number- one micronutrient deficiency.

Some of these investigations are designed to determine how to boost beans' iron bioavailability—the amount of iron our bodies can absorb and use from beans. That might be done in a number of ways, all using plant breeding. One way, of course, would be to increase the level of iron in these legumes. Another approach would be to increase the effects of certain natural compounds that enhance iron bioavailability. A third tactic: decrease the effects of natural compounds that make iron less absorbable.

To discover more about the availability of iron in beans—or in other foods and food components—Glahn developed a laboratory test in 1998 that uses Caco-2 (pronounced KAY-coe) human intestinal cells to give an indication of how our digestive system would treat beans and nutrients from beans.

Glahn says follow-up tests with lab animals "are an important intermediate step between the Caco-2 tests and costly studies with human volunteers." In recent years, Glahn and coresearchers at the ARS Robert W. Holley Center for Agriculture and Health, on the Cornell University campus at Ithaca, have shown that chickens "have promise as an animal model for iron absorption studies."

In research published this year in the journal *Poultry Science,* the scientists report that chickens are sensitive to iron deficiency and that at least a half-dozen different indicators of this deficiency, used in studies with other animals, are also valid for research with chickens.

The team's tests with chickens confirmed their Caco-2 findings, namely, that iron in red beans was less bioavailable to the animals than iron in white beans.

Notes Glahn, "This is the first time this disparity in bean-iron bioavailability has been shown in an animal study. It has implications for human nutrition."

The investigation underscores the contribution that findings from Caco-2 and poultry-based assays might have in helping reverse iron deficiency worldwide.—By Marcia Wood, ARS.

In: Plant Genetic Resources and Food Security ISBN: 978-1-61761-732-4
Editors: Marlena A. Diaz 

*Chapter 6*

# CORN: BOOSTING VITAMIN A LEVELS IN CORN TO FIGHT HUNGER

***Agricultural Research Forum***

Corn is essential to the diets of hundreds of millions of people in developing countries, including those in sub-Saharan Africa But millions of those people are at increased risk of health problems because their corn-based diets lack enough vitamin A. Some 40 million children are afflicted with xerophthalmia, an eye disease that can cause blindness, and 250 million people suffer health problems because of a lack of dietary vitamin A. Agricultural Research Service researchers and colleagues at Purdue University and the International Maize and Wheat Improvement Center (CIMMYT) have made some discoveries that could change that.

Corn contains carotenoids, such as beta-carotene, that our bodies convert to vitamin A, but only a very small percentage of corn varieties have naturally high carotenoid levels. Using genetic and statistical tools, the researchers have identified two genes in corn that are linked to higher beta- carotene levels, and they have developed a cheaper and faster way to screen corn plants for more genes that will produce even higher levels of the essential nutrient. The research is expected to at least triple the levels of carotenoids in Africa's corn and could increase levels in some varieties far beyond that, according to Edward Buckler, a geneticist in ARS's Robert W. Holley Center for Agriculture and Health in Ithaca, New York.

The project, funded in part by the National Science Foundation, included major contributions from geneticist Marilyn Warburton of the ARS Corn Host Plant Resistance Research Unit in Starkville, Mississippi; Torbert Rocheford, a crop geneticist at Purdue University in West Lafayette, Indiana; and Jianbing Yan, from CIMMYT in Mexico.

Corn is one of the world's most genetically diverse food crops. Like people, each ear has a slightly different genetic makeup, resulting in slightly different characteristics. That poses a formidable challenge to scientists trying to understand the genetic basis for any corn nutrient. Direct nutrient-screening techniques are expensive. A common technique, high- performance liquid chromatography, can assess levels of beta-carotene in individual plant lines, but screening a single sample costs $50 to $75. Breeders need to screen hundreds—or more—of plants at a time, making the cost prohibitive. Genetic screening via molecular markers would be a more efficient option for screening large numbers of plants if genes involved in high beta-carotene levels were known. After identification of high-carotenoid corn lines, markers enable efficient transfer of the trait to many new varieties via marker-assisted selection; this is important because farmers in developing nations need high-carotenoid varieties that will grow over a wide range of climates and conditions.

**Research may triple corn's carotenoids.**

## A New Approach and a Better Way to Assay Corn

The team took a new approach to identify specific genes and regions of the corn chromosomes that influence production of carotenoids. They examined the corn genome through "association mapping," a method made possible by recent breakthroughs in statistical analysis and DNA sequencing, techniques that accelerate genetic profiling of crops. Association mapping taps the natural genetic diversity of corn to find new and useful traits.

In their study, the researchers surveyed the genetic sequences of diverse corn from around the world. They found two naturally mutated genes, each producing an enzyme at lower levels than those found in most corn varieties. Plants with either gene mutation have higher levels of beta- carotene, and plants with both mutations have higher levels still. Identification of the two genes using the new methods was an important breakthrough in nutritional plant breeding and has been published in the journals *Science* (2008) and *Nature* Genetics (2010).

After genes are identified via association mapping, markers can be developed from these genes to allow for marker- assisted selection, which is much simpler, faster, and "up to 1,000-fold cheaper" than running the types of chemical tests previously used, Buckler says.

Now, scientists in developing countries can cross the newly identified high beta- carotene lines with local varieties and, applying the markers developed from these two genes, choose progeny that are adapted to local growing conditions but still retain high beta-carotene.

Warburton, Yan, and Michael Gore, a former graduate student in Buckler's lab who is now an ARS geneticist at the U.S. Arid-Land Agricultural Research Center in Maricopa, Arizona, are working with various international organizations, such as CIMMYT, China Agriculture University, and the International Institute for Tropical Agriculture, to train plant breeders in developing countries to use their techniques. Some African maize has as little as 0.1 micrograms of beta-carotene per gram of corn. The researchers are confident they will eventually find genes that result in corn with 15 micrograms of beta-carotene per gram, a target for nutritional scientists around the globe working to improve corn varieties and fight world hunger.

"We see large variation in the corn genome, and that gives us a lot to work with," Warburton says.—By Dennis O'Brien, ARS.

*This research is part of Plant Genetic Resources, Genomics, and Genetic Improvement, an ARS national program (#301) described at* www.nps.ars. usda. gov.

*To reach scientists mentioned in this article, contact Dennis O'Brien, USDAARS Information Staff, 5601 Sunnyside Ave., Beltsville, MD 20705-5129; (301) 504-1624,* dennis.obrien@ars.usda. gov.

Maize varies widely in carotenoid content, which affects the grains' color. The white kernels here have almost no carotenoids, while the orange ones are almost as high in them as carrots. But color does not necessarily indicate beta- carotene levels, so researchers look for beta-carotene differences at the gene level.

*Chapter 7*

# POTATOES: NUTRIENT-PACKED AND PEST-RESISTANT POTATOES FROM ARS RESEARCH

## *Agricultural Research Forum*

Potatoes are America's number one vegetable crop. Per capita, Americans consume about 130 pounds annually. Worldwide, it's the fourth largest crop after wheat, rice, and corn. But it's a wonder that the potato makes it to the dinner table at all, given the myriad pests and diseases that can take hold well before harvest.

There's the Columbia root-knot nematode, which costs U.S. growers $20 million annually; the potato tuber moth; and late blight, which caused the Irish Potato Famine of 1845 and is still responsible for significant losses and control expenses today. Chemical fumigants and fungicides have long been a staple defense for these pests and pathogens. But the onset of resistance in new pest or pathogen biotypes—coupled with environmental concerns about long-term pesticide use—has prompted the search for sustainable solutions in the form of genetic resistance.

## A Recent Defender

Nationwide, ARS researchers are seeking to develop new potato varieties that will not only hold their own against insects and disease, but also maintain their storage quality and deliver nutrients that promote health and well-being in spud lovers the world over.

For example, at ARS's Small Grains and Potato Germplasm Research Unit in Aberdeen, Idaho, geneticist Rich Novy and plant pathologist Jonathan Whitworth spearhead a program to develop new potato lines that are resistant to different biotypes of the late blight pathogen, *Phytophthora infestans*. Toward that end, they're collaborating with Héctor LozoyaSaldaña, a potato researcher in Chapingo, Mexico, where late blight is endemic.

Green plants of the late blight-resistant potato variety Defender, surrounded by susceptible varieties killed by late blight in a test plot at Bonners Ferry, Idaho.

Geneticist Rich Novy harvests tubers of a potato breeding clone at Tetonia, Idaho.

"We send 2,500 breeding clones annually to Chapingo, where Lozoya-Saldaña evaluates them for late blight resistance," says Novy. "We then have a duplicate planting of those same breeding clones at Aberdeen, where—based on his late-blight readings—we concurrently select resistant clones and advance them in our program," based on their agronomic performance under irrigated production in the western United States.

The late blight-resistant cultivar Defender is an example of a recent release (2006) from the program. Defender has helped growers save on fungicides and other expenses associated with controlling late blight, which attacks the crop's leaves and tubers, rendering the latter unmarketable. Over the next few years, Defender may be joined by one more blight-resistant potato variety, depending on how it performs in ongoing trials in Idaho, Oregon, Washington, California, and Texas.

Typically, potatoes evaluated by Novy and Whitworth—and released in collaboration with university colleagues and the grower-supported Potato Variety Management Institute—are selected for their likelihood of success in the western United States. But requests for such releases also originate from other regions of the country and from outside the United States, where some of the same problems occur.

## Saving Stored Spuds

Potato diseases are costly, but so are postharvest losses, which range from 10 to 30 percent of the harvested crop. Postharvest losses result mainly from early sprouting and infections caused by wounds suffered during harvest. Some potato varieties also lose nutritional and processing quality faster than others during extended storage.

"Most potatoes come from family farms that cannot afford to take such losses," says Jeff Suttle, research leader in the ARS Sugarbeet and Potato Research Unit at Fargo, North Dakota, and its work site at East Grand Forks.

Marty Glynn, Suttle's colleague at East Grand Forks, works closely with the Northern Plains Potato Growers Association and public potato breeding programs across the United States to evaluate the storage properties of promising new potato varieties. The evaluations are made using a 1/20-scale processing line that exactly mimics those used by large-scale commercial processors of potato chips and French fries. This collaboration has recently given rise to two named cultivars—Dakota Crisp and Dakota Diamond, which fare well even after 9 months of storage.

Seventy percent of all U.S. potatoes are processed into chips, French fries, and dehydrated flakes. Maintaining adequate potato storage quality for processing—in some cases up to 10 months—is paramount to potato producers and processors.

Two priorities for storage managers are wound-healing and sprout control. Potatoes are wounded during harvest and must heal in order to prevent infection by other pathogens. Chemist Ed Lulai, with ARS in Fargo, has identified hormonal signals stimulating the healing process.

At harvest, potatoes are dormant. During storage, dormancy ends and sprout growth commences. Sprouting, in turn, results in numerous biochemical changes, which diminish the nutritional and processing qualities of potatoes. Postharvest sprouting is typically controlled during storage with chemical inhibitors. The long-term goal of Suttle's program is to find less costly, nonchemical solutions to the problem by identifying the genetic cause for early-sprouting tubers. The researchers have identified internal mechanisms that signal sprouts to grow, and they are currently isolating the genes responsible for these signals. Once identified, these genes can be used in potato breeding programs to modify the sprouting characteristics of any given potato line.

Physiologist Jeffrey Suttle inspects microtubers for signs of sprouting before hormone analysis by mass spectrometry.

Pilot plant operator Dennis Olson (right) and food technologist Marty Glynn prepare slices of potatoes for frying and evaluation.

Improved nutrition is another objective: For example, at Aberdeen, the focus is on elevated protein and vitamin C content. Clearwater and Classic, both varieties released in 2008, boast 30 to 40 percent more protein than the Russet Burbank variety.

At ARS's Vegetable and Forage Crop Research Unit in Prosser, Washington, geneticists Chuck Brown and Roy Navarre are seeking increased antioxidant activity and elevated levels of phytochemicals.

Together with colleagues, they've devised new analytical methods for detecting and measuring phytochemical concentrations in tubers. Using these methods, they found a range of phenolic concentrations—from 100 to more than 1,500 milligrams per 100 grams dry weight—in both wild and cultivated potato lines. Phenolics may help diminish cardiovascular disease, respiratory problems, and certain cancers, the researchers say. One type, chlorogenic acid, is being tested by university cooperators for its potential to lower blood pressure. And, says Navarre, some of these potatoes have high levels of antioxidants (more than 300 micromoles Trolox equivalents per gram dry weight) that rival vegetables like spinach.

Sprouting during storage diminishes a potato's processing and nutritional qualities and means less profit for the producer.

The data from Prosser shows the potential for developing high-phytonutrient potatoes. But without a reliable source of disease and/or pest resistance to protect them, such spuds would be less apt to deliver their health-promoting payloads to consumers worldwide. That's why Brown and colleagues have developed germplasm lines like PA99N82-4, a Russet potato that resists Columbia root-knot nematodes. These wormlike parasites cause unacceptable tuber defects, and fumigants are required to produce potatoes where the nematodes exist.

Each growing region has its own unique combination of pest and pathogen problems or other peculiarities. Thankfully, ARS's potato research locations are strategically located to address them, typically in collaboration with state universities and affected industries. The result is an interconnected network that not only benefits the U.S. potato industry, but other nations as well. That's especially important given today's increasing global commerce and the unique challenges and opportunities it presents.—By Jan Suszkiw and Alfredo Flores, ARS.

*This research is part of Plant Genetic Resources, Genomics, and Genetic Improvement (#301), Plant Biological and Molecular Processes (#302), and Plant Diseases (#303), three ARS national programs described at* www.nps.ars. usda. gov.

*To reach scientists featured in this article, contact Jan Suszkiw, USDAARS Information Staff, 5601 Sunnyside Ave., Beltsville, MD 20705-5129; (301) 504-1630,* jan.suszkiw@ars.usda.gov.

Geneticists Chuck Brown (left) and Roy Navarre examine some of the diverse potato lines prior to analysis of phytonutrients.

Looking for signs of resistance, geneticists Dennis Halterman (left) and Shelley Jansky examine resistant (being held) and susceptible potato plants that have been inoculated with *Phytophthora infestans,* the causal agent of late blight.

Fighting Potato Diseases by Enhancing Germplasm

Geneticists Dennis Halterman and Shelley Jansky, with ARS's Vegetable Crops Research Unit in Madison, Wisconsin, are hunting for wild potatoes that contain resistance to important diseases plaguing potato growers nationwide.

One wild potato Halterman has identified, *Solanum verrucosum,* contains a gene with resistance to late blight. Efforts are under way to cross *S. verrucosum* with cultivated potato and integrate the late blight resistance gene.

The researchers are looking to produce germplasm useful for developing a potato cultivar with resistance to both late blight and early blight, which also affects tomatoes. Early blight, a fungal disease, mainly affects the potato plant's leaves and stems but, if left uncontrolled, can also reduce yield. To create the multi-disease-resistant cultivar, the scientists crossed *S. verrucosum* with another wild potato species that has resistance to early blight, and then crossed the resulting wild potato hybrid with cultivated potato. They currently have seedlings in the greenhouse waiting to be field tested.

Halterman and Jansky are also looking for resistance to *Verticillium* wilt, another fungal disease that can remain in the soil for up to 10 years. Halterman developed a molecular marker to screen germplasm for resistance to this disease, saving the scientists time and effort. They found resistance in the wild potato species *S. chacoense* and produced cultivated potato hybrids that contain the important gene. According to Halterman, this is a good, durable gene that should hold up in the long term.

The scientists are also targeting the potato diseases potato virus Y and common scab.—By Stephanie Yao, ARS.

# Chapter Sources

The following chapters have been previously published:

Chapter 1 – This is an edited, excerpted and augmented edition of a United States Congressional Research Service publication, Report Order Code R41091, dated March 1, 2010.

Chapter 2 – This is an edited, excerpted and augmented edition of a United States Government Department of Agriculture publication, Report Order Code EIB-2, dated May 2005.

Chapter 3 – This is an edited, excerpted and augmented edition of an Agricultural Research publication, dated May/June 2010.

Chapter 4 - This is an edited, excerpted and augmented edition of an Agricultural Research Forum publication, dated May/June 2010.

Chapter 5 – This is an edited, excerpted and augmented edition of an Agricultural Research Forum publication, dated May/June 2010.

Chapter 6 – This is an edited, excerpted and augmented edition of an Agricultural Research Forum publication, dated May/June 2010.

Chapter 7 – This is an edited, excerpted and augmented edition of an Agricultural Research Forum publication, dated May/June 2010.

# INDEX

## D



## E

## F



## G



## H



## I

## J



## K



## L



## M



## N

## O



## P



## Q



## R

## S



## T